OUT OF EGYPT

By Tom Miller

OUT OF EGYPT

COPYRIGHT © 2024 BY Tom Miller

Cover design by Zach Miller Design
ztmillerdesign@gmail.com

ISBN: 9798450771267
Published by: Hemmingway Publishers 2024
Printed In United States

OUT OF EGYPT

A Journey from Bondage to Breakthrough

"I am the Lord your God, who brought you **out of Egypt, out of the land of slavery.**" (Exodus 20:2)

DEDICATION

God has truly blessed me with two amazing, wonderful, and gifted children. Zachary and Mallory are my life! You have brought so much joy to your mother and me over the years. Right from the beginning, you both demonstrated excellence in character, focus, and achieving success. Academically, you both worked hard to achieve honors in High School, and both earned full scholarships and graduated with honors from your respective Universities.

You both have also excelled in athletics, hobbies, and your careers. My children are a gift from God, and I am grateful for the respect and love they have shown to me as their Father. Zach and Mallory, this book is dedicated to you. You motivate me to be a better person, a better Dad, and a better Christian. Thanks for taking it easy on your old man and for cutting me slack when I need it. You're the best, and I love you both to the moon and back!

All My Love,

Pops.

ACKNOWLEDGMENTS

Thank you, Heavenly Father, for instilling within me the desire to write. I am grateful for your leadership and your grace that keeps me going day to day. You are amazing, and I am nothing without you.

Thank you, Michelle, for putting up with me when I write at all hours of the day and night. Thanks for helping me when I get stuck in the middle of a thought, and you bail me out. Thanks for always believing in me and for being my best friend in the whole world!

CONTENTS

PART ONE

LIFE AFTER EGYPT

PART FOUR

FREE AT LAST

INTRODUCTION

Recently, I shared a message with our congregation about the many areas of our lives we have surrendered to the enemy. The sermon's primary point is that so many born-again believers unknowingly give place to the devil. We all have areas that we have not yet surrendered to the Lord, and thankfully, He gives us the grace to let go of them over time. But the real struggle is living in bondage to things we do not even know we are bound by. Some Christians are bound by negative emotions still lingering from childhood, others are bound by ancestral sins from a previous generation, and some are bound by other people, all of which are undetected and kept incognito through the skillful deception of the devil.

This concept is repeated throughout Scripture. Over and over, God's people found themselves entangled with the enemy because of their rebellious nature in worshipping idols, intermarrying with heathen nations, and forgetting God in times of prosperity. Jesus revealed the unconscious bondage of the Pharisees when He told them that if they held to his teaching, they would really be His disciples, then they would know the truth, and the truth would set them free. They answered him and said, we are Abraham's offspring, and we have never been bound to anyone, so how can you say that we will be set free?" The

Pharisees, like so many others, were completely unconscious of their bondage.

After four hundred years of slavery in Egypt, God heard the cries of his desperate children and raised up a deliverer to set His people free. But as difficult as it was for God to get His children out of Egypt, it proved even more difficult to get Egypt out of His children! Having lived in bondage all their lives, the Hebrews knew nothing else but servitude to their earthly taskmasters. Freedom was a lifestyle they had never experienced, and so it was brand new to them. Their immediate shouts of joy and their exuberant cries of celebration as they walked away from Pharoah were quickly met with the stark reality and the hardship of traveling through an extremely hot desert at a sloth-like pace with an entire nation of newly liberated people.

Let's peek into the historical account of when God foretold Moses that his descendants would be held captive by the Egyptians. The scripture reference is Genesis 15:13, "Then the Lord said to him, 'Know for certain that for four hundred years your descendants will be strangers in a country not their own and that they will be enslaved and mistreated there." It was not only prophesied that the Israelites would become the slaves of Pharoah, but God also gave the length of time (430 years) that they would live in bondage and become the servants of their enemy.

The good news is that God is merciful and kind, and He will not leave us in darkness or bondage, but He will deliver us in due time. Verse 14 states, "But I will punish the nation they serve as slaves, and afterward, they will come out with great possessions." And that is exactly what happened: after 430 years of mistreatment and bondage, the Israelites plundered the Egyptians. The Lord caused the Egyptians to look favorably upon His people, and they gave the children of Israel what they asked for. "Every woman is to ask her neighbor and any woman living in her house for articles of silver and gold and for clothing, which you will put on your sons and daughters. And so you will plunder the Egyptians." (Exodus 3:22)

It is just as true for us today. When God delivers us from our captivity, we do not leave empty-handed. Not only does He free us from our taskmasters, but He also provides us with abundant and specific resources to bless us on our journey to freedom and prosperity. Because of His great compassion, like any good parent, God comforts and consoles us after we have suffered a while. "For his anger lasts only a moment, but his favor lasts a lifetime; weeping may stay for the night, but rejoicing comes in the morning." (Psalm 30:5)

As wonderful as it is to be set free, deliverance is just the beginning of a breakthrough. Once our chains are broken and the prison door flings open, the process of freedom just begins. God's part is to supernaturally deliver you, and your part is to choose to live free every day. That means you must first discern and then steer clear of the traps that the enemy has set for you. In other words, just because you have been set free, the devil will not stop trying to bind you up again. In fact, he usually comes back with a vengeance! Jesus taught this principle to the Pharisees: "When an impure spirit comes out of a person, it goes through arid places seeking rest and does not find it. Then it says, 'I will return to the house I left.' When it arrives, it finds the house unoccupied, swept clean, and put in order. Then it goes and takes with it seven other spirits more wicked than itself, and they go in and live there. And the final condition of that person is worse than the first. That is how it will be with this wicked generation." (Matthew 12:43-45)

Out With The Old, In With The New

The common mistake so many people make is that when they get free from the old habits, hang-ups, and hurts from their past, they fail to replace those destructive forces with new reinforcements. When you have been addicted to something for a long time, such as nicotine or alcohol, it feels ecstatic to get delivered from it. A huge burden is lifted, and you feel light on your feet. A new joy emerges from the darkness, and you walk with a new bounce in your step. And although that is a very important step, it is only the first step. The next step is to introduce something new to replace the old. New habits must be formed, new structures implemented, and new and healthy relationships must be established to replace the old. These new additions will not only add value to your life, but they will also become a barrier to block the old destructive forces that kept you in bondage from returning.

While there are many "chains" we may be able to break ourselves, there are certain others that require more faith. Jesus' disciples encountered a particular situation that was above their "pay grade." This was the issue; "a man approached Jesus and knelt before him. "Lord, have mercy on my son," he said. "He has seizures and is suffering greatly. He often falls into the fire or into the water. I brought him to your disciples, but they could not heal him." (Matthew 17:14-16) Jesus demanded that the boy be brought to Him, and he cast out the devil, and he was healed that very moment.

The disciples then asked Jesus, "Why couldn't we drive out the demon?" Jesus responded by saying it was because they didn't have enough faith. One translation says this kind comes out only by prayer and fasting. Some deliverances come easier than others. It depends upon the grip the enemy has on his victim. On several occasions, Delilah tied Samson up with ropes to subdue him and hand him over to the Philistines, but each time, Samson easily broke them. It wasn't until Delilah begged Samson repeatedly, to the point of driving him crazy, that she found out the source of his strength.

After luring him to sleep on her lap, Delilah single-handedly took down the man who killed one thousand Philistines with the jawbone of a donkey! How is it possible for an untrained woman to defeat a known warrior so easily? How could such a feeble person subdue a mighty warrior who was called by God to be a judge over Israel? Samson gave her permission, that's how. The enemy asked, and she received. In this case, Samson, who was driven by his lust for women, allowed the enemy to gain a strong grip on his destiny. So strong that when she bound him this time, he could not break free as he could before.

When Delilah cut Samson's hair, his strength was reduced to that of an ordinary man, and because he was so lost in his passion for Delilah, he had no idea that the Spirit of the Lord had left him. This made it easy for the enemy to bind him and impossible for Samson to break free. Finally, Samson was captured by the Philistines, who gouged out his eyes and took him down to Gaza with bronze shackles. They set him to grinding grain in the prison. So, the man God chose to be a deliverer for his people became a captive instead.

Red Rover, Red Rover

My pastor teaches that there are only two kinds of people in the world: captives and deliverers. You are either bound or free. If you are a deliverer, then you are called to set the captives free, but how can you set anyone free if you yourself are bound? Therefore, it is extremely important for deliverers to stay free from the snares of the enemy. When I was in kindergarten, we played a game called Red Rover. The idea was for two opposing teams to line up facing each other and, while holding hands, call to the other team, "Red rover, red rover, send (insert name) right over. The person invited over was to break through the enemy's grip. If he or she failed to break through, that person was now a part of the other team. However, if they did break through, they could bring someone back to their side. And, of course, the team that captured all the people won.

Whenever they called my name to come over, of course, I would look for the weakest link to break through. I would usually look for a link with two girls or a girl and a smaller boy. I sometimes wonder if the devil invented the game Red Rover because that is exactly how he captures people. He waits until he is invited, usually by our sin or disobedience, and then he pounces on our weakest area for the breakthrough. The Apostle James described it this way, "When tempted, no one should say, "God is tempting me." For God cannot be tempted by evil, nor does he tempt anyone, but each person is tempted when they are dragged away by their own evil desire and enticed. Then, after desire has conceived, it gives birth to sin; and sin, when it is full-grown, gives birth to death." (James 1:13-15)

As you read this book, it is my prayer that you will grasp the importance of having an eternal perspective of your life on this side of heaven. King David said, "But I trust in you, LORD; I say, "You are my God." My times are in your hands; deliver me from the hands of my enemies, from those who pursue me." (Psalm 31:14-15) David saw his life through an eternal perspective, and he realized that God was in control of his time here on earth. David went on to say in Psalm 139:14, "Your eyes saw my unformed body; all my days were written in Your book and ordained for me before one of them came to be."

God's got this...He knew when you took your first breath, and He also knows when you will take your last breath this side of heaven. My challenge for you is to discern God's promptings and to walk in freedom so that you can live in Divine alignment with His will for your life this side of heaven, and as you do, may the blessings of obedience consume you and may the eternal rewards you receive abound in abundance! Once you've tasted true freedom, don't take the bait and walk back into the trap of the enemy. Take the Apostle Paul's advice, "It is for **freedom** that Christ has set us free. Stand firm, then, and do not let yourselves be burdened again by a yoke of slavery." (Galatians 5:1)

Allow me to lead you in prayer as we begin this journey together. "Heavenly Father, I ask you to bring the pages of this book to life like only you can. Use this message to encourage us on our journey from bondage to breakthrough. Grant us discernment to uncover the darkness all around us and to choose to live daily in your perfect light. Holy Spirit, breathe life into these pages, illuminate our minds, and give us a heart of wisdom. Give us grace to receive and to apply the truths you impress upon our hearts, in Jesus' Name, Amen."

PART ONE
LIFE AFTER EGYPT

Out Of Egypt

Chapter One

THIS SIDE OF HEAVEN

"Enjoy life today. Yesterday is gone, and tomor-
row may never come."

-Unknown

It was a beautiful sunny afternoon, with a perfect blue sky, as I floated on my raft in our backyard pool. I gazed up into the sky and saw from 238,900 miles away a brilliant view of the moon. Life took on a new perspective at that moment, as I once again realized just how big our God is and how amazing His creation is. That simple gaze all the way into space catapulted me to a new level of seeing things from an eternal perspective. It made me think about this side of heaven and the total span of my life on this planet.

What is the significance of my time on this side of heaven? What am I doing now that will matter in heaven? That all-important focus, as I gazed upon the moon, reminded me that we only get one shot at this thing called life. Each one of us will only get one opportunity, one lifetime, to do our "thing." The writer of Hebrews puts this in perspective; "Just as people are destined to die once, and after that to face judgment, so Christ was sacrificed once to take away the sins of many, (Hebrews 9:27-28). You live once, and you die once. The whole purpose for the incarnation of Jesus was that His ultimate sacrifice for sinners would provide a way so that we may be restored to God and live with Him for all eternity. Jesus only spent 33 years on this planet but lived His life on purpose, and He accomplished His mission, which was recorded in Luke 4:18-19, "The Spirit of the Lord is on me because he has anointed me to proclaim good news to the poor. He has sent me to proclaim freedom for the prisoners and recovery of sight for the blind, to set the oppressed free, to proclaim the year of the Lord's favor."

The game of golf allows you a mulligan, tennis offers a second serve if you miss the first one, and they say that cats have nine lives, but we as humans only get one golden opportunity to craft a life of significance. That lingering thought reminds me of one of my favorite poems called Only One Life.

Only One Life

BY C.T. STUDD

Two little lines I heard one day,
Traveling along life's busy way.

Bringing conviction to my heart,
And from my mind would not depart.
Only one life 'twill soon be past,
Only what's done for Christ will last.

Only one life, yes, only one,
Soon will its fleeting hours be done.
Then, in 'that day' my Lord to meet,
And stand before His Judgement seat.
Only one life,' twill soon be past,
Only what's done for Christ will last.

Only one life, the still small voice,
Gently pleads for a better choice
Bidding me selfish aims to leave,
And to God's holy will to cleave.
Only one life 'twill soon be past,
Only what's done for Christ will last.

Only one life, a few brief years,
Each with its burdens, hopes, and fears.
Each with its clays I must fulfill,
living for self or in His will.
Only one life 'twill soon be past,
Only what's done for Christ will last.

When this bright world would tempt me sore,
When Satan would a victory score.
When self would seek to have its way,
Then help me Lord with joy to say.
Only one life, 'twill soon be past,
Only what's done for Christ will last.

Give me Father, a purpose deep,
In joy or sorrow Thy Word to keep.
Faithful and true what e'er the strife,
Pleasing Thee in my daily life.
Only one life, 'twill soon be past,
Only what's done for Christ will last.

Oh, let my love with fervor burn,
And from the world now let me turn.
Living for Thee, and Thee alone,
Bringing Thee pleasure on Thy throne.
Only one life, 'twill soon be past,
Only what's done for Christ will last.

Only one life, yes only one,
Now let me say, "Thy will be done"
And when at last I'll hear the call,
I know I'll say "Twas worth it all."
Only one life,' twill soon be past,

Only what's done for Christ will last.

If you are prone to procrastinate, let me encourage you to do everything in your power to overcome the temptation to put off today what you can do tomorrow. James 4:13-14 reminds us, "Come now, you who say, 'Today or tomorrow we will go to this or that city, spend a year there, carry on business, and make a profit.' You do not even know what will happen tomorrow! What is your life? You are a mist that appears for a little while and then vanishes."

We only have one life to live with the same number of hours in each day, the same number of days in each month, and the same number of months in each year, but the real question is…how will you spend that time? Will you live as a captive or as a deliverer? Who and what will be your primary focus? How much time will be wasted with selfish ambitions or unfruitful works? Have you put much thought into the impact you want to make or the legacy you want to leave?

In his book First Things First, author Stephen Covey asks the question, "If you were to pause and think seriously about the 'first things' in your life, three or four that matter most, what would they be? Are these things receiving the care, emphasis, and time you really want to give them?" Do the "first things" in your life matter for eternity? We cannot afford to allow the primary focus of our lives to be temporal in nature. There just isn't enough time for that. We must prioritize our lives with an eternal perspective in view.

The foremost priority of Moses and the nation of Israel was to make it to the Promised Land in Canaan. Their brutal journey through the hot desert and their continual conquest of the existing nations was a daunting task, but they had a goal in mind, and they had a destination to reach. This is what they were promised; "When the Lord your God brings you into the land he swore to your fathers, to Abraham, Isaac and Jacob, to give you—a land with large, flourishing cities you did not build, houses filled with all kinds of good things you did not provide, wells you did not dig, and vineyards and olive groves you did not plant—then when you eat and are satisfied, be careful that you do not forget the Lord, who brought you out of Egypt, out of the land of slavery." (Deuteronomy 6:10-12)

Types & Shadows

A type, a shadow, a pattern, or a figure is a prophetic foretelling of future events. A pre-echo. Like God is so excited about what he is going to do that he drops hints. Like a father preparing his child for what is going to happen. The Exodus was a shadow of things to come, and it is a warning for us to heed. The Apostle Paul explained this in 1 Corinthians 10:1-11. "For I do not want you to be ignorant of the fact, brothers and sisters, that our ancestors were all under the cloud and that they all passed through the sea. They were all baptized into Moses in the cloud and in the sea. They all ate the same spiritual food and drank the same spiritual drink, for they drank from the spiritual rock that accompanied them, and that rock was Christ. Nevertheless, God was not pleased with most of them; their bodies were scattered in the wilderness.

Now, these things occurred as examples to keep us from setting our hearts on evil things as they did. Do not be idolaters, as some of them were as it is written: "The people sat down to eat and drink and got up to indulge in revelry." We should not commit sexual immorality, as some of them did—and in one day, twenty-three thousand of them died. We should not test Christ, as some of them did—and were killed by snakes. And do not grumble, as some of them did—and were killed by the destroying angel. These things happened to them as examples and were written down as warnings for us, on whom the culmination of the ages has come."

Just as our ancestors were born into slavery in Egypt, similarly, we enter this life as slaves to original sin. Sin separates us from God, and there is only one way to remove it, and that is called repentance. The Israelites cried out to God for deliverance, and He heard their cry. He raised up Moses to be their deliverer. Similarly, God raised up His Son, Jesus, to deliver us from our bondage to sin. Romans 6:17-18 reminds us, "But thanks be to God that, though you once were slaves to sin, you wholeheartedly obeyed the form of teaching to which you were committed. You have been set free from sin and have become slaves to righteousness."

No Longer Slaves

Receiving forgiveness for sin doesn't mean that we will never sin again. It means that the power of sin is defeated, and the chains that once held us captive are broken. Now, each of us has a choice to either willfully sin or to abstain from sin. Our old, corrupt nature was prone to sin, and it came naturally to us. But once we are born again, and the Holy Spirit inhabits us, our new nature is no longer prone to sin because it becomes unnatural to us. We take on a new nature which reflects the nature of Christ.

One of the secrets to living sin-free is to daily abide in the Word of God. In John 8:31-33, Jesus confronted the religious leaders concerning their bondage. He said, "If you continue in my word, then you are my disciples; indeed, and ye shall know the truth, and the truth shall make you free." The religious leaders replied by saying that they were not bound to anyone or anything. However, it was their religious pride that kept them ignorant and deluded from the truth. Had they recognized their bondage and confessed their need for freedom, there is no doubt that Jesus would have set them free.

Chapter Two

BARBS & THORNS

"Those who look only to the past or the present are certain to miss the future."

- John F. Kennedy

God never intended for anyone to live in bondage to anything or anyone else, for that matter. He created us to worship Him and live in harmony with one another. His plan for us is not to harm us but to prosper us and to give us a future and hope. God's plan for Israel was to drive out all the inhabitants of the land of Canaan so that His chosen people could dwell there as an inheritance. This is what He told Moses, "Speak to the Israelites and say to them: 'When you cross the Jordan into Canaan, drive out all the inhabitants of the land before you. Destroy all their carved images and their cast idols and demolish all their high places. Take possession of the land

and settle in it, for I have given you the land to possess. (Numbers 33:51-52)

This command was a very strategic strategy designed to keep the Israelites free from the entanglement of the heathen who lived in Canaan. God further told Moses, "'But if you do not drive out the inhabitants of the land, those you allow to remain will become barbs in your eyes and thorns in your sides. They will give you trouble in the land where you will live." (V. 55). Three times, the Lord used the word "all" when He told Moses to rid the land of its inhabitants. No trace of the past can remain if we are to embrace the future God has in store for us. When we are born again, the Bible says that we become a new creation; the old is gone, and the new is here.

Father God knows how easy it is for us to become entangled with the world and its ways, so He tells us plainly in His Word, "Therefore, since we are surrounded by such a great cloud of witnesses, let us throw off everything that hinders and the sin that so easily entangles, and let us run with perseverance the race marked out for us." (Hebrews 12:1-2) John F. Kennedy once said, "Those who look only to the past or the present are certain to miss the future." If we want to embrace all that God has for us and take new territory for the kingdom, we cannot dwell on the past or allow "dead weight" to keep us from reaching our full potential.

Now & Later

One of my favorite treats as a kid was Now and Later candy. If you are not familiar with them, they are bite-sized, fruit-flavored squares that are sort of hard and sort of chewy, and they stick to your teeth when you attempt to chew them. There are so many good flavors, like cherry, grape, and lemon, but to me, the best flavor is watermelon. It not only tastes like watermelon, but it also actually smells like watermelon! I am not sure why they call them now and later, but my best guess is that since they are so difficult to chew, they will last both now and later. Either way, as my memory serves me, since I haven't had one in a long time, they were tasty treats.

Something we cannot fail to lose sight of is the importance of understanding both the temporal aspect (now) and the eternal aspect (later) of our lives. We cannot afford to ignore the certain reality of what comes later while we mindlessly engage in the activities and demands of now. In fact, everything we do now will directly impact a later outcome with eternal consequences. Cultivating an eternal perspective does not mean neglecting our lives in this temporary world. Rather, it shifts our lives into a proper perspective.

The Promised Land provided the Israelites with hope as they journeyed through the vast wilderness with all its challenges and setbacks. And for us, it is the promise of heaven, total paradise that keeps us motivated when life seems out of control. God's Word often refers to the importance of keeping a proper perspective. The Apostle Paul encouraged the Corinthian church, "Though outwardly we are wasting away, yet inwardly we are being renewed day by day. For our light and momentary troubles are achieving for us an eternal glory that far outweighs them all. So we fix our eyes not on what is seen, but on what is unseen, for what is seen is temporary, but what is unseen is eternal." (2 Corinthians 4:16-18)

Earth is a temporary residence for the children of God. This world and all its contents will one day be consumed by fire. So, John's advice is, "Do not love the world or the things in the world. If anyone loves the world, the love of the Father is not in him. For all that is in the world—the desires of the flesh and the desires of the eyes and pride in possessions—is not from the Father but is from the world. And the world is passing away along with its desires, but whoever does the will of God abides forever." (1 John 2:15-17) Jesus gave us a clear standard to gauge where our heart's focus really is; it is where we place our resources that determine whether we are living for this world or the next.

Have you noticed how many storage facilities there are in your city? Where I live, they are building them fast and furious! They are now climate-controlled, multi-story buildings with hundreds of units in various sizes. There are facilities where you can store your R.V. and/or your boat or any vehicles that do not fit in your driveway. I understand that there are many good reasons for these facilities, but I also know that America is obsessed with stuff! Stuff that came into our lives through retail therapy, the need to keep up with the Jones', and/or just because we wanted it. Stuff that won't last, stuff that we cannot take with us to heaven. Stuff primarily intended to satisfy or pacify our desires on this side of heaven.

It is no secret that our society is overcome with materialism. The definition of materialism is a tendency to consider material possessions and physical comfort as more important than spiritual values. Materialism is an indicator that our focus has become bent toward the here and now and has faded from the eternal perspective we had when we first got saved. We were created for heaven, our eternal destination, where we should store our treasures. Scripture warns us, "Do not store up for yourselves treasures on earth, where moths and vermin destroy, and where thieves break in and steal. But store up for yourselves treasures in heaven, where moths and vermin do not destroy, and where thieves do not break in and steal." (Matthew 6:19-20)

The flip side of materialism is living with contentment and being grateful for what you have. The Apostle Paul admonished Timothy, "Now there is great gain in godliness with contentment, for we brought nothing into the world, and we cannot take anything out of the world. But if we have food and clothing, with these, we will be content. But those who desire to be rich fall into temptation, into a snare, into many senseless and harmful desires that plunge people into ruin and destruction. For the love of money is the root of all kinds of evil. It is through this craving that some have wandered away from the faith and pierced themselves with many pangs." (1 Timothy 6:6-10)

Contentment is defined as being in a state of peaceful happiness, satisfied with a certain level of achievement or good fortune, and not wishing for more. It is an attitude of gratitude. Having your needs met is very gratifying, while always striving for more can be extremely tiresome. The Spirit of greed suggests that the more you have, the more you want. The trouble is the devil is also very greedy, and he has his sights set on your stuff.

"The thief comes only to steal and kill and destroy; I have come that they may have life and have it to the full." (John 10:10). What does the thief come to steal, kill, and destroy? Our potential, our time, our opportunities, our stuff, and eventually our hope. This compelling verse of Scripture has both the ability to reveal our personal standing in Christ as well as the power to expose it. We are either being ripped off or living in abundance. The second part of that verse, which says, "I came that they may have life and have it abundantly," indicates that we can have an abundant, bountiful, and blessed life if we can trust God for it.

We always have a choice whether to embrace, dream about, reach for, and live the life that God has promised us...or not. Despite our health challenges, economic conditions, or our current disappointments – we all have the option to believe in the abundant life Jesus offers us. Mark Twain penned the following words for us to consider: "Twenty years from now, you will be more disappointed by the things that you didn't do than by the ones you did do. So, throw off the bowlines. Sail away from the safe harbor. Catch the trade winds in your sails. Explore. Dream. Discover."

I don't know what season of life you are in right now or where you are on your bucket list, but I must ask, what are you waiting for? God's desire for us each day is to live an abundant and purpose-filled life with Him. One day, we will all kneel before Him, either worn out from living life to the fullest or, sadly, with plenty of treads left on the soles of our spiritual shoes. Anticipating that day of accountability encourages me to make the most of every opportunity. I can either make excuses for my failures or embrace the opportunities God gives me; the choice is mine and mine alone to make.

Let's face it, nobody likes to wait on about anything these days. To avoid waiting, we look for shortcuts, self-checkout registers, quick-fix solutions, and get-rich-quick schemes, most of which backfire on us for one reason or another. There are no shortcuts in the kingdom of God. It is in waiting that we gain new strength in which to embrace life's trials with confidence. "But those who wait on the LORD Shall renew their strength; They shall mount up with wings like eagles, they shall run and not be weary, they shall walk and not faint." (Isaiah 40:31)

The book of Hebrews, which lists the heroes of the faith, gives us a proper perspective concerning our lives this side of heaven; "These all died in faith, not having received the promises, but having seen them afar off were assured of them, embraced them and confessed that they were strangers and pilgrims on the earth." (Hebrews 11:13) A stranger is a person who does not know or is not known in a particular community. A pilgrim is a person who travels on long journeys. Neither of these people is either familiar with the place they are in or very close to the people that dwell there. Our Heavenly Father prefers that we do not get too accustomed to life here on earth because this is not our home, spiritually speaking. Earth is our temporary residence, but heaven is our eternal home.

Jesus described our future home this way; "In My Father's house are many mansions; if it were not so, I would have told you. I am going to prepare a place for you. And if I go and prepare a place for you, I will come again and receive you to Myself; that where I am, there you may be also. And where I go, you know, and the way you know." (John 14:2-4) There will be no pilgrims and strangers in heaven, only permanent residents whose names were written in heaven!

The book of Revelation gives us a glimpse of what heaven will be like. "The city had no need of the sun or of the moon to shine in it, for the glory of God illuminated it. The Lamb *is* its light. And the nations of those who are saved shall walk in its light, and the kings of the earth bring their glory and honor into it. Its gates shall not be shut at all by day. There shall be no night there. And they shall bring the glory and the honor of the nations into it. But there shall by no means enter it anything that defiles or causes an abomination or a lie, but only those who are written in the Lamb's Book of Life." (Revelation 21:23-27)

How we conduct our lives now, this side of heaven, will greatly impact our lives later in eternity. First and foremost, we must embrace the gift of salvation freely provided by the blood of Jesus to gain entrance into the kingdom of God. When a religious leader named Nicodemus asked Jesus how to get to heaven, "Jesus answered and said to him, "Most assuredly, I say to you, unless one is born again, he cannot see the kingdom of God." (John 3:3)

You might be asking the same question Nicodemus asked Jesus, "How can someone be born when they are old? "Surely they cannot enter a second time into their mother's womb to be born!" (John 3:4) Jesus was not telling Nicodemus to be born again by earthly means but rather by spiritual means. God is a Spirit, and those who worship Him must worship in Spirit and in truth. Jesus responded to Nicodemus by saying, "Very truly I tell you, no one can enter the kingdom of God unless they are born of water and the Spirit. Flesh gives birth to flesh, but the Spirit gives birth to spirit." (v5-6)

So, how about you? Is your name written in the Lamb's Book of Life? Have you been born again? If not, I invite you to pray a simple but heartfelt prayer asking Jesus to forgive your sins and come into your heart as Lord and Savior of your life. The Bible gives clear instructions on how this works, "If you declare with your mouth, "Jesus is Lord," and believe in your heart that God raised him from the dead, you will be saved. For it is with your heart that you believe and are justified, and it is with your mouth that you profess your faith and are saved." (Romans 10:9-10)

This is sometimes called the prayer of salvation; "Dear Heavenly Father, I know that I have sinned against You, and I have failed to keep Your commands. Please forgive me. I repent of all my sins, and I ask Jesus to come into my heart as my Lord and my Savior. I give you my life because you gave your life for me. Thank you for saving me. In Jesus' Name, I pray, Amen." If you prayed that prayer for the first time, let me be the first to congratulate you on making the best decision of your entire life.

Now that you have received the promise of eternal life, you must develop your relationship with your Heavenly Father through prayer and reading the Bible. The Bible teaches us, like newborn babes, to long for the pure milk of the Word, that by it you may grow in respect to your new salvation. You also need to find a good bible teaching church to help you grow into your newfound faith. This will ensure that you have a solid foundation to stand on as you mature in Christ.

Receiving God's gift of salvation, this side of heaven is the key that opens the door to His eternal Kingdom. In Revelation 3:20, Jesus said, "Here I am! I stand at the door and knock. If anyone hears my voice and opens the door, I will come in and eat with that person, and they will be with me." When we open the door, Jesus comes in, and we now have fellowship with Him. Like any relationship, it takes time and effort to develop, and the good news is that Jesus promises to never leave us nor forsake us...to be with us to the very end!

Perhaps the scene at Golgotha, where Jesus was crucified, paints the most prolific picture of human choice. On either side of Jesus, there hung a criminal who was being punished for his crimes. Unlike Jesus, both men were actually guilty of crimes. While both were receiving the death penalty for their wrongdoings, one of them would receive the ultimate pardon, and the other would be condemned eternally. One of the criminals acknowledged Jesus as the Son of God, and the other did not. One asked Jesus to remember him when He came into His kingdom, and Jesus replied, "Truly I tell you, today you will be with me in paradise." (Luke 23:43)

Both criminals had equal opportunity on this side of heaven to choose for themselves what they believed about Jesus. But that offer expired once they died, and then they were held accountable for their choice. Jesus' offer of paradise has an expiration date; it only exists for those living on this side of heaven. Jesus explained this in a parable about a rich man and a poor man named Lazarus.

"There was a rich man who was dressed in purple and fine linen and lived in luxury every day. At his gate was laid a beggar named Lazarus, covered with sores and longing to eat what fell from the rich man's table. Even the dogs came and licked his sores. "The time came when the beggar died, and the angels carried him to Abraham's side. The rich man also died and was buried. In Hades, where he was in torment, he looked up and saw Abraham far away, with Lazarus by his side. So he called to him, 'Father Abraham, have pity on me and send Lazarus to dip the tip of his finger in water and cool my tongue because I am in agony in this fire.'

"But Abraham replied, 'Son, remember that in your lifetime you received your good things, while Lazarus received bad things, but now he is comforted here, and you are in agony. And besides all this, between us and you, a great chasm has been set in place so that those who want to go from here to you cannot, nor can anyone cross over from there to us.' (Luke 16:19-26)

My desire for you as you read this book is that you would appreciate the gift of life that God has given you this side of heaven. Value it, enjoy it, protect it, and share it as much as you can. Life is short, so live it to the full and squeeze the most out of it that you possibly can. Live with your destiny in focus and set your sights on things yet to come. Remember, the best is yet to come!

Chapter Three

DON'T LOOK BACK

"Burn the bridge; it is either victory or death."

- General George Washington

W hy is it that most of us like to play it safe and take so few risks in life? If you guessed the fear of failure, you are absolutely right! Fear is a terrible weapon of the enemy that paralyzes so many well-meaning Christians. When Jesus appeared to His disciples walking on the water, only Peter was courageously willing to follow His example. "Jesus immediately said to them: 'Take courage! It is I. Don't be afraid.' "Lord, if it's you," Peter replied, "tell me to come to you on the water." 'Come,' he said. Then Peter got out of the boat, walked on the water and came toward Jesus. But when he saw the wind, he was afraid and, beginning to sink, cried out, "Lord, save me!" (Matthew 14:27-30)

Notice how even after Peter mustered up enough courage to get out of the boat and walk a few steps on the water, that fear still raised its ugly head and down went Peter into the water. You may be familiar with this scene. Perhaps you also took a risk and stepped out in faith to do something that required great courage, only to sink in fear and disappointment. That can be extremely difficult to overcome. Ask me how I know. The hardest part is getting back up again and believing that anything is possible, but that is exactly what we must do to overcome our fears and heartaches. Looking back seldom produces good results.

One of my favorite verses in the Bible is Proverbs 24:16, "...for though a righteous man falls seven times, he rises again..." This verse is underlined in my Bible and etched in my heart. To try and fail is not failure. Failure is simply not trying. Be encouraged, my friend; this verse does not indicate that you can only fall seven times. On the contrary, it implies that no matter how many times a righteous person falls, he knows that he can get back up, dust himself or herself off, and get right back in the game.

As a follower of Jesus, you have nothing, nada, zippo, absolutely zero to lose! God is for you, so who can be against you? "If we live, we live for the Lord; and if we die, we die for the Lord. So, whether we live or die, we belong to the Lord" (Romans 14:8). Wow, that is plain amazing! So, what are we afraid of? What are we waiting for to get more spiritual? To get wiser, braver, or stronger? Get real. God is not depending on you or your abilities. He offers you His Spirit instead. "Not by might nor by power, but by my Spirit,' says the LORD Almighty." (Zechariah 4:6).

"Brothers and sisters, think of what you were when you were called. Not many of you were wise by human standards; not many were influential; not many were of noble birth. But God chose the foolish things of the world to shame the wise; God chose the weak things of the world to shame the strong. God chose the lowly things of this world and the despised things – and the things that are not – to nullify the things that are so that no one may boast before him. It is because of him that you are in Christ Jesus, who has become for us wisdom from God – that is, our righteousness, holiness, and redemption. Therefore, as it is written: 'Let the one who boasts boast in the Lord,'" (1 Corinthians 1:26-31).

Did you catch that? God chose you! It is because of Him that you have faith to believe. Can I get a witness? That's right, the God of the universe is not only on our side, but He also chose us to be on His team, and guess what...He never loses! You and I are on the winning team of life, and if we trust fully in Him and keep our eyes on what is ahead, we cannot lose. Sure, there may be some hardships and trials along the way, but at the end of the day, we are overcomers and more than conquerors in Christ Jesus!

What would you attempt to do if you knew that there was no possible way that you would fail? I can think of a bunch of things that I would do, some of them spiritual and some not so spiritual. Some things would have an eternal impact, and other things would not. For example, if I knew that I would not fail, I would share my faith more. I would attempt difficult challenges with confidence. I would focus more on what is possible than what is not. So, why do we shy away from the impossible? I believe it is because we think we have so much to lose. But the truth is, we have nothing to lose.

The fear of rejection is a major deterrent to our potential in Christ. Our past failures inhibit us from our future potential. I have heard people say, "I would be bolder with my witness if I knew that I wouldn't be rejected." I would be great at sales if I knew I wouldn't be turned down so much. Actually, those who work in sales are trained to receive rejection. In fact, they are taught to anticipate much more rejection than acceptance. It usually takes several refusals before you ever get a few sales.

You have probably heard the phrase, "no risk, no reward." While it is rather cliché, it is also very true. You don't have to travel to the casinos in Las Vegas to become a high-stakes gambler, but there are certain risks that can produce tremendous dividends if we dare to take the first step. For example, if you really like someone or even love them, but you are afraid to verbalize your feelings for fear of rejection, you could miss the chance of a lifetime.

Sadly, I have heard of many cases where believers are afraid to share their faith with a family member or a close friend because they are afraid of possible rejection. Later, that person either moved far away or, in some cases, passed away, and the chance to share the gospel was gone. The guilt that results is far worse than the possible rejection they might have encountered. At least if you had shared Christ with them and they rejected you, you would know you did your best, and you would be less likely to feel guilt and shame. More importantly, if any of them actually did receive Christ into their heart, the reward would far outweigh the risk of rejection in the long run.

The fear of failure is also a major stumbling block for many believers. The fear of failing can be immobilizing – it can cause us to wind up doing nothing and, therefore, resist moving forward. We all have different definitions of failure simply because we all have different benchmarks, values, and belief systems. A failure for one person might simply be a great learning experience for someone else. Fear of failure can be linked to many causes. For instance, having critical or unsupportive parents is a cause for some people. Because they were routinely undermined or humiliated in childhood, they carry those negative feelings into adulthood.

Experiencing a traumatic event at some point in your life can also be a cause. For example, say that several years ago, you gave an important presentation in front of a large group, and you did very poorly. The experience might have been so terrible that you became afraid of failing in other things. And you carry that fear even now, years later. Fear of failure can manifest in many ways: A reluctance to try new things or get involved in challenging projects, self-sabotage – for example, procrastination, excessive anxiety, or a failure to follow through with goals. Low self-esteem or self-confidence – commonly using negative statements such as "I'll never be good enough to get that promotion" or "I'm not smart enough to get on that team." Or Perfectionism – A willingness to try only those things that you know you will finish perfectly and successfully.

The first step in overcoming these and any fears is to recognize where it comes from. 2 Timothy 1:7 assures us, "God has not given us a spirit of fear, but of power and of love and of a sound mind." Fear is a tactic of the devil to paralyze the potential of believers. We must recognize the source (the devil) and reject the invitation to fall prey as a victim. I like to define fear with the acrostic, False Evidence Appearing Real.

85 percent of what people worry about never even happens, and with the 15 percent that does happen, 79 percent of people discovered either they could handle the difficulty better than expected or the difficulty taught them a lesson worth learning. This means that 97 percent of what you worry over is not much more than a fearful mind punishing you with exaggerations and misperceptions.

I'll share one of my personal experiences with you: when I was in High School, I had a summer job as a lifeguard at a country club. There was a female lifeguard whose name was Kim, and she was drop-dead gorgeous! She had beautiful blonde hair and, of course, a glistening golden tan and a smile that could melt an iceberg! On top of that, she had a wonderful personality, which made her a perfect 10, at least in my book. Of course, I was secretly in love with her, but there was one really big obstacle standing in my way to true love and bliss...her boyfriend!

Her boyfriend was the quarterback at the University of Iowa and looked like it...if you know what I mean. He was tall, dark, and handsome, and everything else a woman could want in a man. My chances were slim to none, but that did not stop me from admiring her and sometimes flirting with her, of course, when her boyfriend wasn't around. Eventually, I sensed defeat and decided to move on. But little did I know that things were about to change. That Fall, I enrolled in Southern Illinois University, which was also the University where Kim attended. As fate would have it, I ran into her at a party. We talked for a bit, and then she said the most shocking words I had ever heard up to that point in my life...she said, "Do you remember when we were lifeguards at the country club?" I said, "Of course I do." Then she proceeded to tell me that she had the biggest crush on me that Summer, and when I got back up off the floor, I told her that I felt the exact same way.

I could not believe my ears were hearing those words, but it was true. She said it, and my confidence level rose to an all-new high at that moment. I had no earthly idea that she felt that way about me. I remember at that point thinking anything was possible. Never count yourself out of something, no matter how remote your chances are. Just in case you're wondering, my wife gave me permission to tell that story. She knows how blessed I am to be her husband, and we have been happily married for over 30 years. Just saying.

Rear View Mirror

The next time you get in your car, pay attention to the size of your rear-view mirror versus the size of your windshield. In comparison, the rear-view mirror is tiny, and the windshield is huge. Let that be a constant reminder that you need to focus less on what is behind you and more on what is ahead of you. Lessons from the past only serve to keep us from repeating the same mistakes over and over. The Israelites spent forty years wandering in the desert because they focused more on what they could not do than what, with God's help, they could do. The sin of unbelief was prevalent with many of the leaders of Israel.

A typical response in times of trial and hardship is to do an about-face and retreat to the "good ole days." The Israelites encountered extreme difficulty in the desert, so they began to murmur and grumble against Moses and Aaron. They said, "If only we had died by the Lord's hand in Egypt! There we sat around pots of meat and ate all the food we wanted, but you have brought us out into this desert to starve this entire assembly to death." (Exodus 16:3) Over and over, this was the cry of those who could not bring themselves to look through the huge windshield and believe God for a hopeful future. Rather, they chose to focus on the tiny rear-view mirror, looking back to the way things were.

The past can easily become an anchor, holding you back from moving on and taking new ground. While it is good for us to learn from our past, it is important that we do not dwell on our past. A quick glance in the rear-view mirror from time to time is wise, but if you fix your gaze on the rear-view mirror, you will be in great danger of what is happening in front of you. Far too many people sabotage their future by living in their past. It is hazardous to think more about what you cannot change (your past) than about what you can actually do something about (your future). Forward-thinking people tend to find greater success in life than those who spend all their time looking back.

A great illustration of this is found in Genesis chapter nineteen. The cities of Sodom and Gomorrah had become so wicked that God decided to utterly destroy them. Through the faith and prayers of Abraham, God allowed his nephew Lot and his family who lived there to evacuate before He rained down fire and brimstone upon it. God gave specific instructions for Lot and his family: "Flee for your lives! Don't look back, and don't stop anywhere in the plain! Flee to the mountains or you will be swept away." (Genesis 19:17) Lot and his daughters wisely heeded the warning and got out just before the wrath of God fell upon Sodom and Gomorrah, but his wife looked back, and she became a pillar of salt.

What was so important in Sodom and Gomorrah that Lot's wife disobeyed the direct order of the Lord and looked back? What strong ties to either the cities or its inhabitants, or both, could have made her take one last glance? Wasn't the prospect of a better future with her husband and children of greater importance to her? What was she thinking? Were there other family members left behind? Did she forget to bring something with her because she left in haste? The Bible is not clear on why Lot's wife looked back, but one thing is crystal clear...she paid dearly for doing so.

Looking back can prove to be fatal for your career, your dream or vision, your marriage or family, and especially your health and well-being. Jesus said, "No one who puts a hand to the plow and looks back is fit for service in the kingdom of God." (Luke 9:62) C.S. Lewis said, "There are far, far better things ahead than any we leave behind." I remind you to take note of the size of your rearview mirror versus the size of your windshield and keep that information in perspective lest you lose sight of the danger of looking back.

Burn Your Bridges

No matter how wonderful your past or your present life may be, if you do not embrace the potential of your future, you will likely miss some fabulous opportunities. Once such an opportunity came to Elisha. He was plowing with his oxen when Elijah, the prophet of Israel, came and placed his mantle on him. This seemingly insignificant act was a transfer of Elijah's anointing, or his God-given assignment, to Elisha. This was clearly a special moment, an unprecedented opportunity for Elisha, so Elisha quit plowing and then slaughtered his oxen. He burned the plowing equipment to cook the meat and gave it to the people, and they ate together. Elijah then set out to follow Elijah and become his servant.

When General Washington was approaching one of the fiercest battles of his campaign, he and his troops had crossed over a bridge as they were nearing the battle when one of his officers came to him and asked whether to burn the bridge behind them. (It was customary to burn the bridges as they crossed over them to prevent the enemy from slipping up on them from behind & to prevent soldiers from running away from the battle when it became too hot to handle). The great General Washington looked toward the battle and then looked back to where they had come from and then back toward the battle and then gave his answer. "Burn the bridge. It is either victory or death."

Moving On Up

That was a once-in-a-lifetime opportunity for Elisha. He had to make a very important decision, one that required a total change in direction. He had to choose to give up what he had to get what he needed. In most cases, you cannot have both, and this was one of those cases where Elisha had to choose one or the other. He could have held on to the familiar, that which came naturally to him, or choose the unfamiliar, the unknown, and the uncomfortable. We know that he chose the latter, and that was the right choice!

According to leadership expert John Maxwell, "Everything is a tradeoff. Everywhere you leave, you leave some good things. The more successful you are, the harder it is to make tradeoffs. That is why some people become successful and then become flat. Growth demands a temporary surrender of security. It may mean giving up familiar but limiting patterns, safe but unrewarding work, values no longer believed in, and relationships that have lost their meaning." Are you willing to do that? If so, you are a candidate for going to the next level in your field of work or ministry. If not, get comfortable where you are because that is where you will stay until you are willing to make a change.

Maxwell continues, "To gain credibility, you must consistently demonstrate three things:

Initiative – You have to get up to go up.

Sacrifice – You have to give up to go up.

Maturity – You have to grow up to go up.

If you show the way, people will want to follow you. The higher you go, the greater the number of people who will be willing to travel with you." Breakthrough only comes when we are willing to give up whatever has been holding us back. The way forward must become a greater conviction than where you have been. You cannot be in both places at once; you either live in the past or embrace the future.

Letting go is hard to do. That is why so many people are stuck in the past. When what is behind you is greater than what is before you, advancement is nearly impossible. One of the hardest lessons in life is not only letting go but knowing when to let go. Whether it's guilt, anger, love, loss, or betrayal, change is never easy. We fight to hold on, and we fight to let go. "You can't go back and change the beginning, but you can start where you are and change the ending." (Inspiring & Positive Quotes) One of the happiest moments of life is when you find the courage to let go of what you cannot change.

The people of your history are not always the people of your destiny. God uses specific people at specific times in your life to get you to the place He desires you to be. Don't allow your relationships to anchor you to the past. Instead, ask God to send someone to catapult you to another level. When David found himself to be an enemy of King Saul, he really needed a friend. God provided an unlikely person, Saul's son Jonathan, to assist David at this difficult time in his life. This is the biblical account; "As soon as he had finished speaking to Saul, the soul of Jonathan was knit to the soul of David, and Jonathan loved him as his own soul. Saul took him that day and would not let him return to his Father's house. Then Jonathan made a covenant with David because he loved him as his own soul. And Jonathan stripped himself of the robe that was on him and gave it to David, and his armor, and even his sword and his bow and his belt." (1 Samuel 18:1-4)

Jonathan was like a brother to David and proved to be a faithful ally at a difficult time in his life. However, Jonathan died in battle fighting the Philistines. Although Jonathan was a vital part of David's past, he was not present when David was crowned King of Israel. Moses was a part of Joshua's history but not his destiny. Moses led the people to the entrance, but Joshua brought them into the Promised Land. Elijah was a huge part of Elisha's early ministry, but he was taken up into heaven in a miraculous fashion and was absent from his later years of ministry. Be grateful for those God sends to bless you and assist you, but keep in mind the fact that they may be temporary.

Certain people can only take you as far as God allows them to. Keep an eye open for the person God has assigned to help you with a new challenge in your life. Also, make yourself available to God to be like Jonathan was to David to someone he assigns you to. We all need a helping hand to fulfill our destiny, and each of us can become a blessing to others in their time of need.

Out Of Egypt

Chapter Four

BREAKTHROUGH

"In order to have a breakthrough, you gotta have a go-through."

-Joyce Meyer

Freedom is a wonderful blessing, but getting free can be very costly, and staying free can be quite challenging. On the other hand, captivity is a terrible curse that binds and restricts our potential. In the insect world, spiders use a method to capture their prey unlike any other. The spider spins a silky web intended for an unsuspecting creature to pounce on. And, of course, the unsuspecting creature finds out too late that it was drawn into a deadly trap. The victim also realizes that the web is very sticky and that leaving the web is going to be either extremely difficult or totally impossible. The insect then begins to panic and starts to thrash back and forth, which only causes it to become even more ensnared in the web.

Then, the spider wraps its prey in silk to either eat it or store it for a later time. This process can be as quick as forty seconds. The devil works in much the same way. He spins his "web" of wickedness to entice us, tempt us, and draw us close to his trap. If we take the bait and land in his web, we, too, will become "stuck" and paralyzed by his schemes. What happens next is of utmost importance. Either we thrash and squirm, only causing the enemy's grip to become stronger, or we choose to call on God to deliver us. The Apostle Paul and his travelling companion Silas demonstrated the best course of action when we find ourselves bound in captivity.

Acts chapter 16 records the event. After being severely flogged and imprisoned for casting the devil out of a fortune-teller, Paul and Silas were thrown into prison. They were placed in the inner cell, and their feet were fastened in the stocks. Rather than murmuring and complaining as their ancestors did when they came out of Egypt, they chose to pray and sing hymns to God. Imagine your back bleeding and bruised from thirty-nine lashes, you're thrown into a dark inner cell, and your feet are fastened in the stocks. To be honest, would you feel like breaking out in jubilant worship and intercessory prayer? Me neither, but that is exactly what these two men did.

It's important to understand that when your back is against the wall, people are listening to you, and they are watching you to see how you will respond. Oh, by the way, God is also listening and watching to see your reaction. The Bible says that the other prisoners were listening to Paul and Silas pray and worship God. The devil may be able to bind you physically, but he can't steal the praise from your lips! Paul wrote to the church at Corinth, "Though outwardly we are wasting away, yet inwardly we are being renewed day by day." (2 Corinthians 4:6). What took place next was a supernatural miracle. "Suddenly there was such a violent earthquake that the foundations of the prison were shaken. At once all the prison doors flew open, and everyone's chains came loose." (Acts 16:26)

I don't think it was a coincidence that the earthquake happened at that time. I believe that the power of worship and prayer in the midnight hour created a shift in the atmosphere and impacted the Spirit and physical realms. I believe it was a very similar experience when Jesus breathed His last breath upon the cross. The sky grew dark, and thunder and lightning appeared, and the earth shook to the point that people rose out of their graves!

Our response to suffering in times of great difficulty has a direct impact on our destiny, but it also has the power to impact the destiny of those around us. The imprisonment of Paul and Silas resulted in the salvation and baptism of the jailor and his entire family! Not to mention, it resulted in the freedom of all other prisoners. It truly was a supernatural, Holy Ghost jailbreak!

How Breakthrough Comes

Once, while I was in prayer, I heard the words *correctional captivity* in my Spirit. I sensed that it referred to a particular judgment that God would use to restore His people when they wandered away from Him. The Bible gives several instances when the Israelites turned their backs on the Lord and began to embrace the customs of the heathen around them and worship their idols. As a means to correct their rebellion, God would send them into captivity to one of their enemies to give them a wake-up call.

God used many captors, such as the Egyptians, the Babylonians, the Philistines, and the Romans, to rule over His people until they were ready to repent. In this case, the Israelites forfeited their freedom and abused their privileges of liberty. God didn't restrict their freedom; rather, His children took it for granted, and they abused it. They usually remained in bondage until they repented of their rebellious ways and confessed their sins to Jehovah.

The same principles apply to us. If we break the law and if we abuse our privileges of freedom, the law enforcement officer will arrest us, read us our rights and take us to the police station to process us. If they determine it is necessary, they will incarcerate us in a jail cell, and suddenly, we are held captive. Captives have zero options or opportunities. They are at the mercy of their captor, and the first question they usually ask is, "How do I get out of here?" The next thing out of their mouth is usually, "God if you get me out of this, I promise to _____." You fill in the blank.

The breakthrough comes in a variety of ways, and sometimes God revealed the length of the sentence of captivity His children would endure. For example, He told Abraham that his descendants would be enslaved in Egypt for four hundred years. He told Moses that the murmurs would wander in the wilderness for forty years, one year for every day they spied in Canaan. He said, "When seventy years are completed for Babylon, I will come to you and fulfill my good promise to bring you back to this place." (Jeremiah 29:10). Once their sentence was up, God would usually send a deliverer to free His people from their bondage.

At other times, freedom comes through warfare. While David and his men were away fighting for the Philistines, the Amalekites attacked and raided their homes in Ziklag. They burned the city and took the women and everyone else, young and old, captive. David inquired of the Lord, asking if he should pursue the Amalekites and if he would be successful in overtaking them. The Lord affirmed David and told him that he would certainly overtake them and succeed in the rescue.

David took six hundred men, but two hundred of them were too exhausted to fight, so they stayed behind to watch over their supplies. Eventually, David caught up with the Amalekites and overtook them, recovering everything they had stolen. Nothing was missing.

Sometimes, freedom comes from supernatural deliverance. While the church was thriving in Jerusalem, King Herod began to persecute its leaders. He had James put to death by the sword, and he had Peter arrested and put in prison with four squads of soldiers posted to guard him. While Peter was in prison the church was earnestly praying for him. The night before Herod was to bring him to trial, Peter was sleeping between two soldiers bound with chains, and sentries stood guard at the entrance. Suddenly, an angel of the Lord appeared to him and struck him to wake him up. The angel said, "Put on your clothes and sandals, wrap your cloak around you, and follow me." (Acts 12:8)

They supernaturally passed by all the guards and came to the city gate, which opened for them all by itself. Once in the clear, the angel left. Then Peter said, "Now I know without a doubt that the Lord has sent his angel and rescued me from Herod's clutches and from everything the Jewish people were hoping would happen." (v. 11)

Breakthrough sometimes comes in unsuspecting ways. Goliath and the Philistine army had King Saul and the Israelite army bound and paralyzed with fear. For forty days, the armies lined up for battle, and Goliath taunted the Israelites with threats but still no action. Finally, David shows up and asks why no one is responding to this uncircumcised Philistine. Not happy with their response, David decided to do something about it himself. Remember, he was just a child who worked as a shepherd and a messenger boy for his father. He had no military training, no military weapons or protection to engage in battle, and, most importantly, no experience in warfare.

None of that stopped David. He chose to simply use what he was already familiar with to conquer the giant. With his sling and five smooth stones, David ran down the hill toward Goliath, shouting, "You come against me with sword, spear, and javelin, but I come against you in the name of the Lord Almighty, the God of the armies of Israel, whom you have defied. This day, the Lord will deliver you into my hands, and I will strike you down and cut off your head. This very day I will give the carcasses of the Philistine army to the birds and the wild animals, and the whole world will know that there is a God in Israel" (1 Samuel 17:45-46)

And that is exactly what happened. David struck Goliath with a rock from his sling and knocked him out. He then cut off his head, and then the armies of Israel attacked and plundered the Philistines in a great victory. No one saw that coming! How in the world could a teenage shepherd boy defeat a nine-foot giant who was a seasoned warrior and his armor-bearer? It took unwavering faith and absolute courage for David to get his breakthrough.

Another unsuspecting story of deliverance came about through a young lady who went from an insignificant lifestyle as an orphan to becoming the Queen of Persia. The opportunity came about when Vashti, who was the Queen of Persia, refused to attend a banquet in which her husband Xerxes wanted to display her beauty. In his anger, Xerxes banished Vashti from his presence. "If it pleases the king, let him issue a royal decree and let it be written in the laws of Persia and Media, which cannot be repealed, that Vashti is never again to enter the presence of King Xerxes. Also, let the king give her royal position to someone else who is better than she." (Esther 1:19)

Have you ever felt out of place? It's an awkward feeling when you find yourself in an unfamiliar place, and you feel all alone. If you've ever been there, you start asking questions like how did I get here? And then you ask how do I get out of here? Then you make bold declarations that you will never get yourself in a position like that again, ever! While Vashti's position was diminished, Esther's position was about to be elevated. Here's what happened next:

Later when King Xerxes' fury had subsided, he re-membered Vashti and what she had done and what he had decreed about her. Then, the king's personal at-tendants proposed, "Let a search be made for beautiful young virgins for the king. Let the king appoint commis-sioners in every province of his realm to bring all these beautiful young women into the harem at the citadel of Susa. Let them be placed under the care of Hegai, the king's eunuch, who oversees the women, and let beau-ty treatments be given to them. Then let the young woman who pleases the king be queen instead of Vash-ti." This advice appealed to the king, and he followed it." (Esther 2:1-4)

When the king's order was proclaimed throughout the citadel of Susa, many young women were chosen and placed under the care of Hegai, one of whom was Esther. Immediately, she pleased the king and won his favor. He provided her with beauty treatments and special food. He also assigned her to seven female at-tendants selected from the king's palace and moved her and her attendants into the best place in the har-em.

Now, Esther had been raised by her cousin Morde-cai, who was among the Jews who were exiled by Neb-uchadnezzar, king of Babylon, and she had not revealed her nationality and family background because Mordecai had forbidden her to do so. "The king was attracted to Esther more than to any of the other women, and she won his favor and approval more than any of the other virgins. So he set a royal crown on her head and made her queen instead of Vashti. And the king gave a great banquet, Esther's banquet, for all his nobles and officials. He proclaimed a holiday throughout the provinces and distributed gifts with royal liberality." (Esther 2:17-18)

After these events, King Xerxes honored Haman the Agagite, he elevated him and gave him a seat of honor higher than that of all the other nobles. All the royal officials at the king's gate knelt and paid honor to Haman because the king had commanded this. However, Mordecai would not kneel or pay him honor. When Haman saw that Mordecai would not kneel or pay him honor, he was enraged. When he learned that Mordecai's people were Jews, he scorned the idea of killing only Mordecai. Instead, Haman looked for a way to destroy all of Mordecai's people throughout the whole kingdom of Xerxes.

Haman plotted to annihilate the entire population of Jews by telling the king lies about them. He told Xerxes that the Jews practiced offensive customs and did not obey the king's laws. Haman suggested that a decree be made by the king to destroy the Jews, and he offered the king money to sweeten the deal. The king told Haman to keep the money and do to the people what you wished. The trap was set, and the bait was taken. All that was left was to execute the plan, but God had a different plan in mind for His people. When the enemy threatens and takes His people captive, God raises up a deliverer to save the day.

God raised up Esther for such a time as this. He gave her the courage to confront the king about Haman's evil plot, and the king was enraged when he discovered what was in the heart of his noble servant. So, the king ordered that Haman be hung on the very gallows that he built for Mordecai. What the devil means for harm, God turns around for our good. It is important for born-again Christians to understand their place in Christ. They are ordained to be the head and not the tail, above and not beneath.

"And God raised us up with Christ and seated us with him in the heavenly realms in Christ Jesus, in order that in the coming ages he might show the incomparable riches of his grace, expressed in his kindness to us in Christ Jesus," (Ephesians 2:4-7). We are much more effective when we operate from our place of authority. The seventy disciples came back with a glowing report after Jesus had sent them out to minister. They said, "Lord, even the demons submit to us in your name" (Luke 10:17).

Because of Jesus, we have an advantage in the world, much like a race car driver who wins the pole position. Pole position is the place at the inside of the front row at the start of a racing event. This position is typically given to the vehicle and driver with the best qualifying time in the trials before the race. In Christ, we have been given the upper hand. We have a tremendous vantage point that grants us greater opportunity in life. That is why God wants us to see ourselves "in Christ" and to focus on our peculiar position, not our past limited condition.

Our position in Christ represents our personal standing with God: we are either guilty or not guilty, debtors or free, enemies or friends. Understanding our position makes a difference in how we live day in and day out. In His goodness and by His grace, God changes our standing, or our position, before Him when we place our trust in Christ for the forgiveness of our sins. Ephesians 2:7 affirms that our new status flows out of that wonderful phrase, "the riches of God's grace."

Did you know that every believer has been given an eternal position before God? Because God has "reconciled us to Himself." We are no longer His enemies. We are now considered His friends. II Corinthians 5:17–18 reads, "Therefore if any man is in Christ, he is a new creature; the old things passed away; behold, new things have come. Now all these things are from God, who reconciled us to Himself through Christ and gave us the ministry of reconciliation." Our old position has been exchanged for a new one. We are a new creation. God made it. Our spiritual birth placed us in our new position. We are now identified with Jesus Christ. This is true of every believer.

There is a difference between our spiritual position in Christ and our present condition. For example, haven't you ever asked yourself why, as a believer in Christ, you need to ask forgiveness for your sins, but yet the Bible says before God's eyes you are as righteous and as sinless as Christ because He sees you in Christ. How can both be true?

Our being as righteous as Christ is our position in Christ, not our condition now. The position is about our legal status. When the Bible says that by faith in Christ, we are justified (declared not guilty), it is referring to our position. The biblical term 'justified' in reference to Christians has solely a judicial meaning. Judicially, we believers in Christ are declared to be not guilty, but in our condition, we still have guilt because we still have sin.

Our position in Christ on this side of heaven determines whether we will experience a life of abundance or settle for mundane mediocrity. After confessing that the thief comes to steal, kill, and destroy, Jesus declared that He came to give us life and life more abundantly, see John 10:10. What is the thief most likely to steal? Not only our position but our potential as well. When we walk in our full authority in Christ, we are pregnant with potential. The Bible makes it clear that nothing is impossible for those who believe, and that is an enormous threat to the thief.

As a child, I learned how important the pecking order is. I am the youngest of eleven children. I am familiar with the bottom of the barrel. I lived at the wrong end of the food chain and had to learn survival skills at an early age. Hand-me-downs were a regular part of my lifestyle. It was difficult to be seen and heard and much safer to blend into the background when things got hairy. I rarely felt like I had an advantage being the baby of the family, except when my siblings accused me of being the favorite child. I lived in the shadow of ten older siblings; some shadows cast a favorable light, and others a not-so-favorable light. But I am proud of my upbringing, even though my parents divorced when I was in Jr. High, and finances were tight most all the time. Amidst the common battles all families face, love and respect were the foundation of our home. Nevertheless, I always felt like I was starting in the back of the pack.

The Advantage of Favor

God's favor, or grace, is God giving us the ability to do something which is humanly impossible for us to do. For example, it is only by God's undeserved favor that we can experience eternal life, and it is only by God's grace that we have the ability to live for the Lord. "For you bless the righteous, O Lord; you cover him with favor as with a shield." (Psalm 5:2) The "favor of God" can be described as a divine kindness or an act of true compassion on the part of God Himself toward needy and undeserving hu-man recipients. Often, in Scripture, this act of God toward unworthy men or women is referred to as God's "grace"—which means "the unmerited favor of God,"

God's favor is totally and unequivocally undeserved and unmerited. There is nothing we can do to earn or merit his favor. He supplies us with His favor or grace to-tally at His initiative and only because of His love for us. "For the Lord God is a sun and shield; the Lord bestows favor and honor. No good thing does he withhold from those who walk uprightly." (Psalm 84:11)

God's favor is greater than any earthly treasure be-cause the treasures of this world will one day perish by fire. God's favor is eternal, and it is not limited to time and space as worldly treasures are. Favor will take you further than money ever will because while there are limitations to currency, there are no limitations to God's favor. The psalmist reminds us, "May the favor of the Lord our God rest on us; establish the work of our hands for us— yes, establish the work of our hands." (Psalm 90:17)

Recognize the Favor of God

Even when you are in an environment that is hostile, God's favor is upon you, and you will be distinguished. In Genesis 39:4, the Bible says: "And Joseph found grace in his sight and served him: and he made him overseer over his house, and all that he had he put into his hand." Favor can also be referred to as exceptional kindness and privilege from God. Divine favor is a privilege that God bestows upon His children; we all have it, but not all of us recognize it.

Every born-again Christian has the blessing of God's grace upon their lives, but not every believer walks in the grace of God. By recognizing the favor of God in our lives, the breakthrough is inevitable because we realize that God is for us and not against us. Satan's goal is to keep believers ignorant of their God-given potential by hiding it from them. If you haven't already done so, I pray that you will discover the amazing grace of God and the gift of favor in your life to accomplish the supernatural.

PART TWO
SOBER MINDED

Chapter Five

SOBER MINDED

"So then, let us not be like others who are asleep, but let us be awake and sober."

1 Thessalonians 5:6

W hat is a sober mind? 1 Peter 4:7 gives us a picture of a sober mind; "The end of all things is at hand; therefore, be self-controlled and sober minded for the sake of your prayers. The Bible defines "sober" as having a disciplined, controlled, sound mind. To have a sober mind is to be in control of your thoughts, emotions, and actions. It means to be circumspect, aware, and alert to the enemy.

The term sober minded literally means "free from intoxicating influences. "More broadly, being sober-minded means that we do not allow ourselves to be captivated by any type of influence that would lead us away from sound judgment. The sober-minded believer is not "intoxicated" by the world system and is, therefore, calm under pressure, self-controlled in all areas, and rational. Those who are sober-minded will be alert to the need to pray and will do so at opportune times.

Unfortunately, we see the opposite of sober mindedness displayed regularly in our modern culture. Sinfulness, irresponsible choices, foolish experimentation with harmful substances or behaviors, and crude joking are among the practices of the intoxicated.

Sober Defined:

Serious, sensible, and solemn: marked by earnestly thoughtful character or demeanor. Unhurried, calm, marked by temperance, moderation, or seriousness. Showing no excessive or extreme qualities of fancy, emotion, or prejudice. If substances control us, we cannot also be controlled by the Holy Spirit. Sober-minded believers choose to abstain from practices that would lead them into sin.

Your emotions follow where your mind goes, and that drives your behavior. So, if you learn how to be sober-minded, then you can be a steady steward and a solid witness of the gospel. 1 Peter 1:13, "Therefore, preparing your minds for action, and being sober-minded, set your hope fully on the grace that will be brought to you at the revelation of Jesus Christ."

Being sober-minded does not mean living a sour, joyless existence. Rather, sober-minded Christians are to be continually filled with the joy of the Holy Spirit. "For the kingdom of God is not eating and drinking, but righteousness and peace and joy in the Holy Spirit." Romans 14:17 Eliminating foolishness, frivolity, and mind-numbing worldliness from our lives allows us to focus on what is real, eternal, and inspiring. "Finally, brothers and sisters, whatever is true, whatever is noble, whatever is right, whatever is pure, whatever is lovely, whatever is admirable—if anything is excellent or praiseworthy—think about such things. "Philippians 4:8

In these last days, there is an urgency that compels us to live sober-minded lives. "The night is almost gone, and the day is near. Therefore, let us lay aside the deeds of darkness and put on the armor of light. Let us behave properly as in the day, not in carousing and drunkenness, not in sexual promiscuity and sensuality, not in strife and jealousy. But put on the Lord Jesus Christ and make no provision for the flesh in regard to its lusts. "Romans 13:12-14

Benefits of a Sober Mind

- You will have a greater knowledge of God.

"This is the covenant I will make with the people of Israel after that time," declares the LORD. "I will put my law in their minds and write it on their hearts. I will be their God, and they will be my people. No longer will they teach their neighbor, or say to one another, 'Know the LORD,' because they will all know me, from the least of them to the greatest," declares the LORD. "For I will forgive their wickedness and will remember their sins no more." Jeremiah 31:33-34

- You will have a stronger faith.

"So then, just as you received Christ Jesus as Lord, continue to live your lives in him, rooted and built up in him, strengthened in the faith as you were taught, and overflowing with thankfulness. See to it that no one takes you captive through hollow and deceptive philosophy, which depends on human tradition and the elemental spiritual forces of this world rather than on Christ." Colossians 2:6-8

- You will be grounded in truth.

"And you, who once were alienated and hostile in mind, doing evil deeds, he has now reconciled in his body of flesh by his death, in order to present you holy and blameless and above reproach before him, if indeed you continue in the faith, stable and steadfast, not shifting from the hope of the gospel that you heard, which has been proclaimed in all creation under heaven, and of which I, Paul, became a minister." Colossians 1:21-23

- You will have Biblical discernment.

"Beloved, do not believe every spirit, but test the spirits to see whether they are from God, for many false prophets have gone out into the world." 1 John 4:1

- You will stay steadfast in ministry.

"Therefore, my beloved brothers, be steadfast, immovable, always abounding in the work of the Lord, knowing that in the Lord your labor is not in vain." 1 Corinthians 15:58

- You will suffer well through trials and tribulations.

"Blessed is the man who remains steadfast under trial, for when he has stood the test, he will receive the crown of life, which God has promised to those who love him." James 1:12

- You will receive protection from false doctrine.

"Indeed, all who desire to live a godly life in Christ Jesus will be persecuted, while evil people and impostors will go on from bad to worse, deceiving and being deceived. But as for you, continue in what you have learned and have firmly believed, knowing from whom you learned it." 2 Timothy 3:12-14

- You will be able to resist temptation.

"No temptation has overtaken you that is not common to man. God is faithful, and he will not let you be tempted beyond your ability, but with the temptation, he will also provide the way of escape, that you may be able to endure it." 1 Corinthians 10:13

- You will have a guarded heart.

"Above all else, guard your heart, for everything you do flows from it." Proverbs 4:23

- You will walk in humility.

"As a prisoner for the Lord, then, I urge you to live a life worthy of the calling you have received. Be completely humble and gentle; be patient, bearing with one another in love. Make every effort to keep the unity of the Spirit through the bond of peace." Ephesians 4:1-3

Steps of Sobriety

- Pause For Perspective.

The first step to a sober mind is to PAUSE and put the situation in perspective. Life is full of stresses that stir up our emotions and attitudes, but when we stop and get the proper perspective, we can respond in a way that glorifies God. Take a time-out and allow the Holy Spirit to rule over your flesh and your feelings. Proverbs 16:32 says, "Better a patient person than a warrior, one with self-control than one who takes a city."

- Arrest Your Thoughts.

Put the cuffs on your brain! A sober mind is a con-trolled, disciplined, sound mind. To train our carnal minds to be sober, we must arrest our thinking and take our thoughts captive. 2 Corinthians 10:5 says, "We demolish arguments and every pretension that sets itself up against the knowledge of God, and we take captive every thought to make it obedient to Christ."

- Speak The Truth.

When the world, the flesh, and the devil bombard you with lies, submit yourself to biblical truth. Stop and ask yourself, "What is true?" because the truth changes everything. The more truth we flood our minds with, the less susceptible we are to believe lies. Every scripture you commit to memory is a deposit into your spiritual storehouse. You can't withdraw what you don't have, so be intentional to memorize scripture.

John 8:31-32, "To the Jews who had believed him, Jesus said, "If you hold to my teaching, you are really my disciples. Then you will know the truth, and the truth will set you free."

- Know Your Enemy.

Remember that difficult people and trials and tribulations are you, not your enemy...the devil is! Chaos and contention originate with him. Ephesians 6:12, "For our struggle is not against flesh and blood, but against the rulers, against the authorities, against the powers of this dark world and against the spiritual forces of evil in the heavenly realms."

If you want to be cool in the crazy and calm in the chaos - cleave to Jesus and leave the drama, the distraction, and the divisive devices of the devil behind.

- Walk In the Spirit.

When we yield to the Spirit's power, our responses to conflict and contention will greatly improve. Galatians 5:16-17, "So I say, walk by the Spirit, and you will not gratify the desires of the flesh. For the flesh desires what is contrary to the Spirit, and the Spirit what is contrary to the flesh. They are in conflict with each other, so you are not to do whatever you want."

Romans 8:6-7, "The mind governed by the flesh is death, but the mind governed by the Spirit is life and peace. The mind governed by the flesh is hostile to God; it does not submit to God's law, nor can it do so."

- Pray Without Ceasing.

Prayer is essential for a sober mind. Pray for a scripture to help in time of need, for deliverance from the enemy, for wisdom in handling situations, and for strength to endure hardships. Make God your first choice, not your last chance! 1 Thessalonians 5:17 exhorts us to "Pray without ceasing." To have a sober mind, we must constantly take our cares and concerns to the Lord in prayer.

Philippians 4:6-7, "Do not be anxious about anything, but in every situation, by prayer and petition, with thanksgiving, present your requests to God. And the peace of God, which transcends all understanding, will guard your hearts and your minds in Christ Jesus."

Spiritual Sobriety

One translation of the word "sober" is the Greek word sophronizo, which translates as sound mind, discipline, or correct. The root word sophron means safe, sound mind, self-controlled, moderate as to opinion or passion, and temperate. An example of the use of the word sophron is found in 1 Peter 4:7, which says, "The end of all things is near. Therefore, be alert and of sober mind so that you may pray."

Therefore, the biblical command here is to have a self-controlled mind so that we can pray effectively. If our minds are intoxicated with doubt and unbelief, it will greatly impact the potential of our prayers. Another translation of "sober" is the Greek word, grēgoreō, which translates to keep watch, be on guard: awake, of sober mind, stay awake, to wake up.

An example of the word Gregorio is found in 1 Thessalonians 5:6-8, "So then, let us not be like others, who are asleep, but let us be awake and sober. For those who sleep, sleep at night, and those who get drunk, get drunk at night. But since we belong to the day, let us be sober, putting on faith and love as a breastplate and the hope of salvation as a helmet." Gregorio implies watchfulness against sin and temptation. Several scriptures support this implication: Matthew 26:41, "Watch and pray so that you will not fall into temptation. The spirit is willing, but the flesh is weak."

1 Corinthians 10:12, "So, if you think you are standing firm, be careful that you don't fall!"

1 Corinthians 16:13, "Be on your guard; stand firm in the faith; be courageous; be strong."

Colossians 4:2, "Devote yourselves to prayer, being watchful and thankful."

The following scriptures pertain to another level of watching, which is to be a watchman:

Psalm 127:1, "Unless the Lord builds the house, the builders labor in vain. Unless the Lord watches over the city, the guards stand watch in vain."

Jeremiah 31:6, "There will be a day when watchmen cry out on the hills of Ephraim, 'Come, let us go up to Zion, to the Lord our God.'"

Acts 20:29-31, "I know that after I leave, savage wolves will come in among you and will not spare the flock. Even from your own number, men will arise and distort the truth in order to draw away disciples after them. So be on your guard! Remember that for three years, I never stopped warning each of you night and day with tears."

The verb form of the word "sober" is nepho, and it signifies being free from the influence of intoxicants. An example of the use of nepho is found in 1 Thessalonians 5:6, "So then let us not sleep, as others do, but let us keep awake and be sober." Jesus told us that He would return like a thief in the night and when we would least expect Him to come. That is precisely why we cannot afford to be snoozing when He returns. Sober means awake and alert, fully cognizant and prepared for something. I want to be in that state of mind as much as possible, looking for the return of Christ.

Chapter Six

SOBER JUDGMENT

"Just as love was the motivation for the crucifixion, so wrath is the motivation for divine judgment."

- T. C. Miller

R omans 12:3, "For by the grace given to me I say to everyone among you not to think of himself more highly than he ought to think, but to think with sober judgment, each according to the measure of faith that God has assigned." Sober judgment means that we see ourselves from God's perspective and the world from His vantage point as well. Christ first, others second, and me third, is God's way for us to think. Self-centeredness is the enemy to avoid.

While it is true that we are fallen and must struggle with sin throughout our lives, we must not allow this truth to overshadow the fact that we are made in the image of God and are "being renewed day by day." Titus 2:2, "Older men are to be sober-minded, dignified, self-controlled, sound in faith, in love, and in steadfastness."

2 Timothy 4:5, "As for you, always be sober-minded, endure suffering, do the work of an evangelist, fulfill your ministry."

1 Thessalonians 5:6-8, "So then let us not sleep, as others do, but let us keep awake and be sober. For those who sleep, sleep at night, and those who get drunk, are drunk at night. But since we belong to the day, let us be sober, having put on the breastplate of faith and love, and for a helmet the hope of salvation."

1 Timothy 3:2-3, "Therefore an overseer must be above reproach, the husband of one wife, sober-minded, self-controlled, respectable, hospitable, able to teach, not a drunkard, not violent but gentle, not quarrelsome, not a lover of money."

Romans 12:2, "Do not conform to the pattern of this world but be transformed by the renewing of your mind. Then you will be able to test and approve what God's will is—his good, pleasing, and perfect will."

2 Timothy 1:7, "For God has not given us a spirit of fear, but of power and of love and of a sound mind."

Spiritual Intoxication

People can be drunk in many different ways, not only through wine or strong drink. For example, we can be drunk with silliness, ambition, passion, fame, grief, life, desire, vanity, success, or power. In scripture, God often likens false doctrines and teachings to wine because they cause people to lack sound judgment, impairing their ability to think clearly or act righteously.

If you've ever been intoxicated from drinking wine or strong drinks, then the parallels between the mind under the influence (of Satan) and being spiritually drunk make perfect sense. Because, like wine, sin may make you feel good for the moment, but ultimately it has severe consequences. By contrast, spiritual sobriety is the state of the mind ruled by Christ's spirit, and it brings eternal pleasure in the truth, as opposed to what makes us feel good for the moment.

Those who choose to drink the wine of spiritual corruption may have joy for the moment, but the wrath of God abides upon them. So, drunkenness, as it relates to Christianity, is often ambiguous language, painting a portrait of the consequences brought about by sin. Revelation 17:1-2, "One of the seven angels who had the seven bowls came and said to me, "Come, I will show you the punishment of the great prostitute, who sits by many waters. With her, the kings of the earth committed adultery, and the inhabitants of the earth were intoxicated with the wine of her adulteries.""

This is referring to Babylon. A prostitute is one who is married, has been false to her husband's bed, and has forsaken the guide of her youth and broken the covenant of God. She had been a prostitute to the kings of the earth, whom she had intoxicated with the wine of her fornication. This wine that makes kings drunk is the seduction of false teachings. In the spirit of disobedience, denial, and delusion, they are impaired, unable to make sound judg-ments. 2 Peter 2:1-3, "But there were also false prophets among the people, just as there will be false teachers among you.

They will secretly introduce destructive heresies, even denying the sovereign Lord who bought them—bringing swift destruction on themselves. Many will follow their depraved conduct and will bring the way of truth into disrepute. In their greed, these teachers will exploit you with fabricated stories. Their condemnation has long been hanging over them, and their destruction has not been sleeping."

v. 13-14, "Their idea of pleasure is to carouse in broad daylight. They are blots and blemishes, reveling in their pleasures while they feast with you. With eyes full of adultery, they never stop sinning; they seduce the unsta-ble; they are experts in greed—an accursed brood!"

v. 17-18, "These people are springs without water and mists driven by a storm. The blackest darkness is re-served for them. For they mouth empty, boastful words and, by appealing to the lustful desires of the flesh, they entice people who are just escaping from those who live in error."

v. 20-21, "If they have escaped the corruption of the world by knowing our Lord and Savior Jesus Christ and are again entangled in it and are overcome, they are worse off at the end than they were at the beginning. It would have been better for them not to have known the way of righteousness than to have known it and then to turn their backs on the sacred command that was passed on to them."

Those caught up with Babylon become drunk—spiritually drunk—as a result of imbibing its way of life. Figuratively, wine has significant spiritual meaning. When wine is used to symbolize doctrine, it is "Old Wine" or fermented grape juice. That is false doctrine. "New Wine," or unfermented grape juice, represents true doctrine. Wine, along with grain and oil, represented God's covenant blessings promised to Israel for obedience. We also see God withholding those necessities for disobedience.

Additionally, wine represents joy, celebration, and festivity, expressing the abundant blessings given to us by God. A drunk's mind becomes dizzy, fuzzy, and unfocused. His perception of reality changes, becoming distorted and uncertain. His body staggers under the effect of the drug, not reacting normally as the drinker commands it to act. At the same time, he is deluded into thinking he has greater power than before becoming drunk. The reality is that he has made himself a helpless victim and is dangerous to himself and others.

The wine in this word picture of Revelation 17:2 is Babylon's way of life. In Revelation 18:2, the wrath is the penalty that comes down upon its hapless victims as they practice the sins of their unfaithfulness to God. Fornication figuratively portrays faithlessness, such as one would experience within a covenant relationship such as marriage.

A Depraved Mind

Romans 1:18-32, "For the wrath of God is revealed from heaven against all ungodliness and unrighteousness of people who suppress the truth in unrighteousness, because that which is known about God is evident within them; for God made it evident to them. For since the creation of the world, His invisible attributes, that is, His eternal power and divine nature, have been clearly perceived, being understood by what has been made, so that they are without excuse."

"For even though they knew God, they did not honor Him as God or give thanks, but they became futile in their reasonings, and their senseless hearts were darkened. Claiming to be wise, they became fools, and they exchanged the glory of the incorruptible God for an image in the form of corruptible mankind, of birds, four-footed animals, and crawling creatures. Therefore, God gave them up to vile impurity in the lusts of their hearts so that their bodies would be dishonored among them."

"For they exchanged the truth of God for falsehood and worshiped and served the creature rather than the Creator, who is blessed forever. Amen. For this reason, God gave them over to degrading passions, for their women exchanged natural relations for that which is contrary to nature, and likewise, the men, too, abandoned natural relations with women and burned in their desire toward one another, males with males committing shameful acts and receiving in their own persons the due penalty of their error."

"And just as they did not see fit to acknowledge God, God gave them up to a depraved mind, to do those things that are not proper, people having been filled with all unrighteousness, wickedness, greed, and evil; full of envy, murder, strife, deceit, and malice; they are gossips, slanderers, haters of God, insolent, arrogant, boastful, inventors of evil, disobedient to parents, without understanding, untrustworthy, unfeeling, and unmerciful; and although they know the ordinance of God, that those who practice such things are worthy of death, they not only do the same but also approve of those who practice them."

The condition of a depraved mind is characterized by an inherent deficiency of moral sense and integrity. It consists of evil, corrupt, and perverted intent, which is devoid of regard for human dignity and which is indifferent to human life. Those who reject this true knowledge of God are given to a "depraved mind." The term means "failing a test." In the NT, it is used of an athlete being "disqualified" (1 Cor. 9:27), of false profession of faith (2 Tim. 3.8), of being "unfit and worthless" (Titus 1:16), and to thorns and weeds (Heb. 6:8). It was also of metals that were rejected by refiners because of impurities.

The person who rejects the quest for deeper knowledge of God is, in fact, rejected by God and becomes spiritually worthless, unqualified, and unfit for righteousness. To forfeit the privilege of seeking God is to put something or someone before God. That is idolatry! Sadly, there is not only an anti-God attitude in the modern world but also a critical attitude against Him and His people. People spend their time in political movements that attempt to silence God and His ministers of grace while pursuing their own ends.

Characteristics of a Depraved Mind

- Suppressing Truth with Wickedness.

The wrath of God is being revealed from heaven against all the godlessness and wickedness of people who suppress the truth by their wickedness – Romans 1:18. This person has knowledge of the truth but chooses to reject it by their wicked actions.

- Willfully Rejecting God.

For although they knew God, they neither glorified him as God nor gave thanks to him – Romans 1:21. The person with a depraved mind is aware of God's existence; they just choose to operate as if he doesn't exist.

- Offering Redirected Worship.

They exchanged the truth about God for a lie and worshiped and served created things rather than the Creator – Romans 1:2. This person has switched their object of affection. In their foolishness, they choose to worship the things they see, which are created, instead of the one who made them, who is the Creator.

- Having No Limits or Restrictions on Sinful Behavior.

They have become filled with every kind of wickedness, evil, greed, and depravity. – Romans 1:29. The person who possesses a depraved mind has taken the restrictions off their sinful boundaries. Nothing is off-limits.

- Demonstrating No Desire for Change.

Although they know God's righteous decree that those who do such things deserve death, they not only continue to do these very things but also approve of those who practice them. – Romans 1:32. This person will not only hold on and continue in their wicked behavior, but they will also encourage others to follow suit. In fact, they will applaud those who follow after them in their degradation.

Depraved defined:

Corrupt, debased, debauched, decadent, degenerate, degraded, demoralized, perverse, perverted, reprobate, sick, unclean, unwholesome, warped. Romans 1:18-32 shows us the necessity of the gospel. If sin didn't make God angry, there would be no need for Jesus to die on the cross. If God is only love, then He could simply decide not to punish any sins. But this passage shows us that God has holy anger.

This isn't the kind of anger we might have if someone cheats on us, lies to us, or violates our "rights." It's more like the righteous anger we might feel if we heard about a person abusing children or cheating the elderly. This wrath might be demonstrated in many ways. It is shown in the way God made up the world to punish those who violate his moral order. It is more clearly seen by God's direct intervention in the world, such as the flood, Sodom and Gomorrah, the sentence passed on Adam and Eve, etc.

Just as love was the motivation for the crucifixion, so wrath is the motivation for divine judgment. Notice here the object of God's wrath. It is the ungodliness and sinfulness of man. It is the sin that makes God angry. God's holy anger should teach us to respect God and realize He is not only a loving Father, but He is also a just Father who will punish those who don't comply with His moral laws. No sin escapes God's attention. There is no doubt we live in a perverse generation with warped and unwholesome people in positions of leadership. While those people may stir up holy anger inside of us and make us want to react in anger, we must leave that job to God. He is our vindicator, and judgment belongs to Him.

Our role is to pray for those people and to keep our minds focused on Jesus and His Word. As you know, the mind is a battlefield, and warfare is constant. Therefore, we should absorb the words of Isaiah 26:3, "You will keep in perfect peace those whose minds are steadfast because they trust in you."

Chapter Seven

RENEW YOUR MIND

"Most of all, let the Word of God fill and renew your mind every day. When our minds are on Christ, Satan has little room to maneuver."

- Billy Graham.

R omans 12:2, "Do not conform to the pattern of this world but be transformed by the renewing of your mind. Then you will be able to test and approve what God's will is—his good, pleasing, and perfect will." What Does 'Renewing of the Mind' Mean? Simply stated, renewing your mind according to Romans 12:2 means interpreting life through the lens of God's Word and the inspiration of the Holy Spirit rather than through the lens of your experience, woundedness, trauma, preferences, or the opinions of others. It's a fundamental shift toward seeing the world, yourself, others, God, and especially what's possible from a Kingdom perspective. It's making a daily, moment-by-moment choice to choose the Mind of

Christ, which lives inside of us as new creations rather than operating from our soulish mind the way we did before we were saved.

Why Does the Bible Emphasize the 'Renewing of the Mind'? Unless Christians learn to renew their mind, they will continue to walk in defeat, struggle, and confusion as they desire to experience a Spirit-led life. This is the predicament of most believers who live lives of religious obedience and obligation yet are void of any real Spirit-led power. Without renewing your mind, the only two options are to wait, hope, and beg God to change you or to work, sweat, and strive on your own to achieve the results you so long to experience. Neither is God's best.

Renewing your mind aligns your mind with the truth of God's Word by learning to recognize the lies of the enemy, replace them with the truth of God's Word, and then reinforce that truth every time the enemy comes at them with those same lies. In 2 Corinthians 10:5, God's Word teaches us that "We destroy arguments and every lofty opinion raised against the knowledge of God and take every thought captive to obey Christ..." We will talk more about strongholds in chapter 9. Taking a thought captive literally means to capture or conquer that thought as you identify it and compare it to God's Word. Does this thought agree with God's promises over my life or not? If not, then I cast it away—rejecting its influence in my life – and I plant in its place one of God's promises from His Word.

In 2 Peter 1:3-4, the Bible says, "His divine power has given us everything we need for a godly life through our knowledge of him who called us by his own glory and goodness. Through these, He has given us His precious and magnificent promises, so that through them you may become partakers of the divine nature, now that you have escaped the corruption in the world caused by evil desires." Did you get that? It's through God's precious promises that we participate in the divine nature.

Renewed thoughts resulting in new beliefs help you engage with God's plan for your life because thoughts fuel your beliefs. Godly kingdom-focused thoughts are great fuel for your godly beliefs, while ungodly, fear-driven lies and half-truths fuel ungodly beliefs. Renewing your mind is not just a spiritual process but a physiological one as well. When you intentionally change the way you think to align with God's Word, it creates new connections and pathways in your brain to make that process easier and more preferred over time.

Through the process of neuroplasticity, your brain can be reconfigured to align with the truth of God's Word and thus create the solutions, strategies, and opportunities that best align with God's plan for your life. The Word of God confirms this in Proverbs 23:7 KJV, "For as he thinketh in his heart, so is he." Your thoughts (which ultimately create your beliefs, resulting in your actions) create the boundaries of your life. If you want to experience a different life—the abundant life Jesus promised—you must learn to intentionally renew your mind.

Ways To Renew Your Mind

• Recognize: Intentionally "capture" each thought that comes into your mind and compare it to God's Word and His promises for your life. Ask yourself, "Does this thought that I'm having right now reflect what God's Word says about this situation, or does it reflect a lie that I know doesn't represent God's heart for me?"

• Replace: If the thoughts you're having don't agree with God's Word, then replace them with promises from God's Word. It can be as simple as asking the Holy Spirit to bring to your remembrance a scripture that is opposite of the lie you may be faced with currently. You can also just do a quick internet search for scripture verses based on the lie you're trying to replace. Remember, thoughts are like seeds, and whatever you allow to be planted in the garden of your heart will bring forth a harvest in your life. Choose your seeds wisely.

• Reinforce: Every time you are flooded with those same types of thoughts that don't agree with God's best for your life, simply reject and replace them with God's Word. One way you can accelerate this process is to write down your newly created Biblical affirmations on note cards, laminate them, and carry them with you for easy access. You can also post them in places that you see often, like your office wall, your desk, your car, or even your bathroom mirror. This can feel tedious at first, but as you do so in the context of gratitude, faith, and expectancy, God's presence will infuse this process, making it become more and more natural.

• Visualize: God created you with an incredible imagination to see, sense, feel, and create new realities in your mind before they ever come into the physical realm. It's one of the primary ways we co-labor with the Holy Spirit to release His Kingdom in and through our lives. As you are replacing and reinforcing new thoughts based on God's Word, imagine in your mind's eye what it would look like and feel like to experience life within this new reality. This further stimulates your brain and accelerates your own ability to come into agreement with God's best for your life.

• Affirm: Create affirmation statements based on God's Word that, along with your visualization exercises, continue to reinforce the truth of God's Word inside your heart. Over time, your subconscious mind will become convinced of the new truth you're aligning your heart with rather than the old way you've been living.

You can absolutely live the abundant life Jesus promised if you master the process of renewing your mind, including recognizing the lies of the enemy, intentionally replacing them with the truth of God's Word, and reinforcing that truth through visualization and affirmations. As you do, you'll be amazed at the double doors of favor that seem to instantaneously appear in your life simply because you came into agreement with God's design for your life rather than trying to make it happen all on your own.

The Problem with Our Minds

There are many who think that the only problem with the human mind is that it doesn't have access to all the knowledge it needs. So, education becomes the great instrument of redemption — personal and social. The Bible has a far more profound analysis of the problem. In Ephesians 4:23, Paul uses a striking phrase to parallel Romans 12:2. He says, "Be renewed in the spirit of your minds." The spirit of your mind means that the human mind is not a sophisticated computer managing data, which it then faithfully presents to the heart for appropriate emotional responses.

The mind has a "spirit." In other words, our mind has what we call a "mindset." It doesn't just have a view; it has a viewpoint. It doesn't just have the power to perceive and detect; it also has a posture, a demeanor, a bearing, an attitude, and a bent. "Be renewed in the spirit of your mind." The problem with our minds is not merely that we are finite but that we are fallen.

Our mind has a spirit, a bent, a mindset that is hostile to the absolute supremacy of God. Romans 8:6-7, "For to be carnally minded is death; but to be spiritually minded is life and peace. Because the carnal mind is enmity against God: for it is not subject to the law of God, neither indeed can be." Renewing our minds is important because it allows us to set our thoughts and affections on things above instead of on earthly things (Colossians 3:2). Too often, our minds are consumed with worry, anxiety, fear, and stress about the things of this world, which we cannot control. But when we renew our minds in Christ, we can focus on the things that are truly important – like our relationship with God and others and what His Word says.

Renewing our minds is also very important because it helps us to better understand and apply the Word of God

(Psalm 119:11). When our minds are filled with the truth of God's Word, we are better equipped to make wise decisions and live in a way that pleases God. Daily renewing our minds enables us to stand firm against the enemy's lies and schemes (Ephesians 6:11-12). The devil will try to discourage us, deceive us, and lure us away from God and His will for our lives. But when we have our minds renewed in Christ, we can resist his lies and remain faithful to God.

Here are some more ways we can renew our minds in Christ: we can pray for God to transform our mind according to Philippians 4:8, we can fill our mind with the Word of God according to Joshua 1:8, and we can memorize and meditate on Scripture throughout the day according to Psalm 1:2. Renewing our minds is not an option, it is a must if we are to live effective and productive lives in Christ.

Put off the Old, Put on the New

Ephesians 4:22-24 admonishes us to "Put off your old self, which belongs to your former manner of life and is corrupt through deceitful desires, and to be renewed in the spirit of your minds, and to put on the new self, created after the likeness of God in true righteousness and holiness." Once we are born-again, we take on an entirely new nature. The old carnal nature is replaced with a brand-new spiritual nature. 2 Corinthians 5:17 describes it this way, "Therefore if anyone is in Christ, he is a new creation. The old has passed away; behold, the new has come."

You are not a better version of the former you. You are now a totally new and different person. You are new in heart and new in spirit (Ezekiel 36:26). You are made of a different substance now unrelated to your old sinful nature. As Adam was made in the likeness of God but later fell, so Christ was made in the likeness of sinful flesh to redeem us from the fall, that by grace, through faith, a new creation might be brought into being. If you are a child of God, you are a unique individual, fearfully and wonderfully made with purpose and destiny instilled within you. Therefore, leave the old nature in the dust and pursue with all your heart the new nature which was imparted to you by the goodness of God.

Chapter Eight

THE MIND OF CHRIST

"Sow a thought, and you reap an action; sow an act, and you reap a habit; sow a habit, and you reap a character; sow a character, and you reap a destiny."

- Ralph Waldo Emerson

It's very important to keep the main thing, the main thing. Broken focus destroys vision and potential. Distractions and disappointments derail our hopes and dreams and discourage our once-optimistic hearts from pressing in. We thrive only when we keep our eyes on the prize. I recall a particular wrestling tournament that I entered one summer when I was in college. It was a freestyle tournament held at Saint Louis University in St. Louis, Missouri. I was extremely determined to win this tournament and I considered any opponent an absolute obstacle in my way on the path to victory. One by one, I defeated my opponents, but I didn't just beat them. I pinned most of them quickly. I guess you could say I had the eye of the tiger! Mentally, I pictured myself beating each opponent before the match ever began, and with that tenacious focus, I brought home the tournament title along with the gold medal.

After winning that tournament in style, I asked myself why I didn't wrestle like that all the time. The answer is that I had never been that focused on a match before. All I was thinking about during that tournament was winning, and every opponent that I faced was trying to keep me from the title. I kept the main thing the main thing. Imagine what could happen if every Christian had that kind of focus. What would the world look like if we were able to continually have the mind of Christ and keep the main thing the main thing?

The Apostle Paul asked, "Who has known the mind of the Lord so as to instruct him? But we have the mind of Christ." (2 Corinthians 2:16). Part of the struggle in keeping our focus is understanding times and seasons. King Solomon was an incredibly wise man. He wrote in Ecclesiastes 3:1, "There is a time for everything and a season for every activity under the heavens." Anyone can look at their watch or a clock on the wall and tell you what time it is. That is what is called Chronos time. Chronos measures time in seconds, minutes, hours, days, years, etc. This is the word that measures the quantity of time. There are some English words, like Chronological, which take inspiration from the same word.

Kairos measures time in the best moments of life. It measures the quality of time. It does not measure minutes, but it measures moments. Kairos is translated as "the right time" from Ancient Greek. It refers to a "spiritual opportunity" from a Christian theological standpoint. That wrestling tournament was a Kairos moment for me and even a defining moment because it birthed a new confidence within me. I discovered how effective I could be if I set my mind to do something and kept my focus unshaken.

We all live life in times and seasons. According to King Solomon, there is "a time to be born and a time to die, a time to plant and a time to uproot, a time to kill and a time to heal, a time to tear down and a time to build, a time to weep and a time to laugh a time to mourn and a time to dance, a time to scatter stones and a time to gather them, a time to embrace and a time to refrain from embracing, a time to search and a time to give up, a time to keep and a time to throw away, a time to tear and a time to mend." a time to be silent and a time to speak, a time to love and a time to hate, a time for war and a time for peace," (Ecclesiastes 3:2-8).

The effectiveness of our lives can either be measured in Chronos time, which represents the temporal/carnal side of life, or in Kairos time, which represents the eternal/spiritual side of our lives. The Bible says in Psalm 90:10, "Our days may come to seventy years, or eighty if our strength endures; yet the best of them are but trouble and sorrow, for they quickly pass, and we fly away." Chronos represents the natural realm, while Kairos represents the supernatural realm. On the day the Lord gave the Israelites victory over the Amorites, Joshua prayed to the Lord in front of all the people of Israel. So, the sun stood still, and the moon stayed in place until the nation of Israel had defeated its enemies.

Elijah prayed for the rain to stop, and no rain fell for three and a half years. Then Elijah prayed, and it began to rain again. These men were able to dramatically impact Chronos' time through specific, powerful Kairos moments. Their spiritual prayers impacted natural time and created a supernatural result. Some believers get discouraged because they feel like they have run out of time to be useful to the Lord. While that may be true in Chronos' time, it's not necessarily true in Kairo's time.

Abraham's wife Sarah was ninety years old when she gave birth to their son Isaac. Moses was eighty years old when he began the arduous Exodus of the Israelites from Egypt. With God, a day is like one thousand years, and a thousand years are like a day. In the blink of an eye, God can do what it would take a lifetime for us to do. On this side of heaven, we need all the Kairos moments we can get. The task before us to reach the world with the gospel is enormous, but with God, all things are possible!

In order for us to pursue the mind of Christ, we must commit to walking in the Spirit. The Spirit searches all things, and He knows our very thoughts. While we do not know the thoughts of God, the Spirit reveals them to us as we walk in alignment with the Word of God. Those who choose not to walk in the Spirit do not understand the things of the Spirit because they are spiritually discerned. In fact, they consider them foolishness and foreign to them. But for those who walk in the Spirit, we have the mind of Christ.

The carnal nature keeps us in bondage to the flesh, which translates to a limited potential of understanding. Sin and darkness go hand-in-hand, and both are a result of prideful rebellion to God and His commands. One of the greatest enemies of life in the Spirit is the reliance upon reason and logic. The book of Proverbs warns us that there is a way that seems right to a man, but in the end, it leads to death. If God intended for us to figure things out on our own and depend upon our own understanding, we wouldn't need Him. One of my favorite scriptures is "Trust in the Lord with all your heart and lean not on your own understanding; in all your ways submit to Him, and He will make your paths straight." (Proverbs 3:5-6)

One of the most difficult things in life to do is to trust God in blind faith. The Israelites were promised a land flowing with milk and honey, but only those who developed trust in a God they couldn't see face-to-face were able to experience it. Doubt and unbelief are the result of leaning heavily upon reason and logic to guide us. Biblical faith makes no logical sense. Faith is the substance of things hoped for and the evidence of things not seen. Try telling that to a judge in a courtroom where they're looking for hard evidence and solid facts!

Perfect Peace

The carnal mind is an enemy of God. It is not subject to Him, nor can it be according to scripture. Romans 8 tells us," The mind governed by the flesh is death, but the mind governed by the Spirit is life and peace." (v. 6). If we choose to live in the realm of the flesh, we cannot please God. We must believe that God exists and that He rewards those who earnestly seek Him because, without faith, it is impossible to please God. True freedom can only come about by living a life of faith and trust in God and His Word.

Isaiah 26:3 gives us insight into how we can live in perfect peace; "You keep him in perfect peace whose mind is stayed on you because he trusts in you." Trusting in God at all times is the key to living in His perfect peace. The treacherous desert and the lack of daily provisions proved to be a trust killer for the children of Israel. Even though God provided several miracles and demonstrations of His power, the extenuating circumstances of the unknown and the uncertainty of the future caused their ability to trust in Jehovah to diminish and disappear.

The very center of the Bible is Psalm 118:8, and that verse says, "It is better to trust in the Lord than to put confidence in man." At the very core of His Word, God is asking us to trust Him more than we trust others. Even with his best intentions, man will fail because of his inherent fallen nature. The only one who will never fail us is Jesus. "Love never fails. But where there are prophecies, they will cease; where there are tongues, they will be stilled; where there is knowledge, it will pass away." (1 Corinthians 13:8). The reason that prophecies and tongues and wisdom cease is that they are the works of man, but God is love and love always was and always will be.

Joyce Meyer states in her article entitled The Way to Live with the Mind of Christ, "We need to get all the junk in our minds out of our way so we can keep running our race in Christ Jesus and have the victory God wants us to have. Then, we are ready for action to follow God's plan for our lives. If you commit to setting your mind on God's Word, renewing your mind with truth, and getting stinkin' thinkin' out of your way, then you will experience the fullness of new life that we can all have in Christ. All it takes is a little more progress, one day at a time."

The mind of Christ is not something God reserves for "perfect" Christians. All believers have access to the mind of Christ through faith. However, we also still have the old mind and all its deceptive ways of dealing with it. We may, at times of weakness, give in to the lure of sin. We may also be hindered by false doctrines or even choose to use our freedom in unhealthy and unproductive ways. Even the most seasoned believers struggle with their thoughts because the mind is a spiritual battlefield.

Researchers indicate that We think between 50,000-70,000 thoughts per day, which means between 35 and 48 thoughts per minute per person. Our minds are constantly working non-stop like a computer sorting data. Certainly, not all those thoughts are good or healthy thoughts that generate wellness and peace. A large portion of our thoughts are inspired by the devil, whose intentions are clear, "The thief comes only to steal and kill and destroy." (John 10:10). Another large portion of our thoughts are focused on our immediate circumstances pertaining to our health, family, work, finances, etc.

So, how much of our thought life is dedicated to our spiritual well-being? As I mentioned earlier, Proverbs 23:7 indicates, "For as he thinks in his heart, so is he." This means that our hearts are directed by our thoughts. While we may not always feel secure in our salvation or confident in our walk with God, it is important to remember that we walk by faith and not by sight. In other words, we cannot allow our feelings to dictate our spiritual course of direction. Jesus promised to never leave us and to never forsake us, yet there are times in the life of every believer when we feel destitute and all alone.

It is critical at those very times when we feel isolated, discouraged, or fearful that we meditate on the Word of God. Psalm 42:11 provides a good remedy for poor thoughts, "Why are you cast down, O my soul? And why are you disquieted within me? Hope in God; For I shall yet praise Him, The help of my countenance and my God." Another great scripture that has encouraged me many times when my emotions are under siege from the enemy is Proverbs 3:5-6, "Trust in the Lord with all your heart and lean not on your own understanding; in all your ways submit to him, and he will make your paths straight."

The battle for spiritual freedom is fought in the mind. Therefore, it is imperative that we constantly examine our thoughts. Our enemy is subtle and seductive. Satan is relentless when it comes to corrupting the way we think. It has been said that our thoughts determine our decisions, our decisions determine our actions, and our actions determine our habits. So, what we do habitually starts with a thought, and Satan knows that if he can successfully plant an errant thought, it is possible to change our behavior.

Ralph Waldo Emerson said, "Sow a thought, and you reap an action; sow an act, and you reap a habit; sow a habit, and you reap a character; sow a character, and you reap a destiny." Our destiny is tied directly to and starts with our own thoughts. In the Bible, Paul reminds us how important our thoughts are. In Philippians 4:8, he writes,

"Finally, brothers and sisters, whatever is true, whatever is noble, whatever is right, whatever is pure, whatever is lovely, whatever is admirable — if anything is excellent or praiseworthy — think about such things."

The Devil Made Me Do It

Remember hearing that old phrase, "The devil made me do it?" The truth is, the devil can't make you do anything, but he sure can tempt you to. No believer is exempt from the temptations urged by our adversary. We are all tempted, and we are all given an opportunity to be led astray if we are not careful. When tempted, no one should say, "God is tempting me." For God cannot be tempted by evil, nor does he tempt anyone, but each person is tempted when they are dragged away by their own evil desire and enticed. Then, after desire has conceived, it gives birth to sin; and sin, when it is full-grown, gives birth to death." (James 1:13-15)

Satan tempted Jesus many times while in the desert, but each and every time He was tempted, Jesus quoted scripture to combat and overpower the devil. You and I must approach temptation in the same way. We, too, must use the sword of the Spirit, which is the Word of God, to combat the enemy. One of my favorite scriptures in the Bible is 1 Corinthians 10:13, "No temptation has overtaken you that is not common to man. God is faithful, and he will not let you be tempted beyond your ability, but with the temptation, he will also provide the way of escape, that you may be able to endure it."

Staying free requires a disciplined thought life. Managing our emotions begins with guarding and choosing our thoughts. Your mind is not a playground for the enemy, nor is it a community field where he can sow seeds of corruption as he pleases. We must guard our minds and protect our thoughts. Then and only then can we have the mind of Christ.

PART THREE

STRONGHOLDS

Out Of Egypt

Chapter Nine

STRONGHOLDS

"Before the truth can set you free, you need to recognize which lie is holding you hostage."

- Unknown

Strongholds are faulty thinking patterns that are based on lies and deception. Deception is one of the primary weapons of the devil because it is the foundation for strongholds. One of the most common and potentially devastating strongholds is an incorrect image in your mind of God and how He sees us. Those who see God as a cruel taskmaster tend to live their lives in an unhealthy fear of God. They find it difficult to feel God's love and presence and to walk in His mercy and grace.

A quite common stronghold is yielding to an incorrect and unhealthy view of ourselves. If the devil can keep you from knowing who you are in Christ and the potential you have according to the Word of God, then he has a successful stronghold over you. This stronghold keeps believers in bondage to guilt, shame, and lack of self-esteem. It is Satan's goal to cause believers to question their salvation, their place in the kingdom, and their value to the body of Christ.

This type of stronghold in the Bible refers to negative patterns of thinking, prideful thoughts, or worldly messages that have written negative messages on the minds and hearts of believers. These types of thoughts, over time, can build powerful spiritual strongholds that need to be torn down with the truth of God's Word.

The greatest trick the devil ever pulled was convincing the world he didn't exist. Satan and his demons would rather deceive people in relative silence by shooting constant accusatory arrows against our hearts and our minds so that we feel so beat up emotionally that we barely have the capacity to ponder the truth of God's word, which says the exact opposite of all the enemy says.

How Do Strongholds Form?

Strongholds often start with a wound we experience, a hurt or disappointment that makes our heart fertile ground for seeds of lies to be planted by the enemy. On this foundation, the devil then begins to build, brick by brick, a wall of lies, inaccurate ideas about the person of God, erroneous interpretations of Scripture, prideful thoughts, and distorted perceptions of how God sees us and feels about us when we sin.

For example, if someone that you trusted, who should've played a father-figure role in your life, deeply betrayed your trust and hurt you, and as a result, the enemy lied to you and told you that no one could be trusted and you are on your own. You then become fiercely independent, so much so that you have trust issues, which has then resulted in your struggling to trust God's promises in your life! When we buy into these lies and agree with the enemy, thereby ceding control, that area of our life becomes a stronghold, a fortress that the enemy vigorously defends and retains control of as long as he can. The spirit of offense is a mighty stronghold primarily because it is so often justified by the offended. Here's the real issue: if you can be offended, you will be offended. The scripture warns us, "Offenses will certainly come, but woe to the one through whom they come!" (Luke 17:1)

There are many ways that a stronghold is formed. Let's take a look at some of the tactics the enemy uses to bind us up.

Unconfessed Sin

The Bible teaches that there is no sin in your life so great that the blood of Jesus isn't greater than and hasn't already been paid for. However, what we willingly choose to keep hidden will remain hidden until we bring it to light or God shines His light on it. The spiritual enemy of your soul thrives on secrecy and will establish a stronghold through any unexposed sin in your life.

Confess all known active sins in your life and repent of them. Be specific with God in your prayers and lay in raw detail all that you thought and said that led to your sinful action. As you do, take great comfort in this promise from your Father in heaven - 1 John 1:9 says, "If we confess our sins, he is faithful and just and will forgive us our sins and purify us from all unrighteousness." Repentance means that you change your mind and intentionally live a life that is different from the sinful pattern you used to engage in.

Occult Activity

Whether it's Voo Doo, Wicca, Fortune Telling, Black or White Magic, Spiritism, Ouija boards, conjuring spirits, or even yoga (depending on who your focus is on during meditations), it's like placing a "vacancy" sign on your life for evil spirits. Have you knowingly or unknowingly dabbled in any occult-like practices in your past? That may be where you opened the door to the enemy.

First, renounce any commitments or oaths you made to any of those groups and declare your allegiance to Jesus Christ as the sole Lord and Savior of your life. Then, plead the blood of Jesus over your mind, your heart, and your spirit. As you do, remember that Jesus Christ is now seated at God's right hand in heavenly *places,* above all principality and power and might and dominion. In other words, the Spirit of Jesus who lives in you and to whom your new allegiance belongs is greater and more powerful than the evil presence behind what you just renounced.

Lies You Have Believed

These include lies about God, the Bible, about yourself, and lies about how God sees you. Lies that the enemy has been playing over and over in your mind for years, which, in your pain, you've unfortunately bought into. What lies have you believed?

Reject those lies the devil tries to whisper in your heart about your self-worth or the Father's love for you. The Holy Spirit of God may bring conviction or heaviness over sin in your life, but the purpose is to lead you back into a right relationship with God. If what you're "hearing" brings you shame or guilt, it's condemnation, and it's not of God and must be rejected. In Philippians 4:8, we're told what to focus on instead of lies, "Finally, brothers, whatever is true, whatever is honorable, whatever is just, whatever is pure, whatever is lovely, whatever is commendable, if there is any excellence, if there is anything worthy of praise, think about these things."

Unforgiveness

If you are still nursing bitterness against an individual or a group of people, the enemy will take advantage of that to keep you in bondage. The people who wounded you may not deserve to be forgiven, but we forgive because Jesus Christ forgave us for an even greater sin against God. We must choose to forgive because it sets us free from bondage produced by bitterness.

How many times should we forgive? Seventy times seven times if necessary! Consider also that when you forgive someone, it's not that you're allowing them to "get away" with anything, but rather, it's you trusting that God is wise enough to handle justice on your behalf.

Generational Curses

You may have picked up on the fact that certain unhealthy behavioral and thought patterns present in your life were also present in your parents' lives and/or your grandparents' lives as well. Sometimes, unhealthy, even ungodly family traditions and family patterns are passed down from generation to generation until someone finally stands up and says, in Jesus' Name, the buck stops here!

You must renounce all affiliation with the sins passed down to you. As Joshua did, draw a line in the sand and declare, "But if serving the Lord seems undesirable to you, then choose for yourselves this day whom you will serve, whether the gods your ancestors served beyond the Euphrates, or the gods of the Amorites, in whose land you are living. But as for me and my household, we will serve the Lord.: (Joshua 24:15)

Living in the Past

Margaret Storm Jameson, the English author, once expressed the view that we all spend too much time living in the past, feeling regret for lost joys or shame for things badly done. Even when our minds turn to the future, she said, we spend an inordinate amount of time longing for it or dreading it. "The only way to live," she said, "is to accept each minute as an unrepeatable miracle . . . Work at your work. Play at your play. Shed your tears. Enjoy your laughter. Now is the time of your life."

In his book, "No Wonder They Call Him Savior," Max Lucado shares, "Not many second chances exist in the world today... Just ask the kid who didn't make the little league team, the fellow who got the pink slip, or the mother of three who got dumped for a 'pretty little thing.' Not many second chances. Nowadays, it's more like, "It's now or never," "Around here, we don't tolerate incompetence" Gotta get tough to get along." "Not much room at the top." "Three strikes and you're out."

Does that sound familiar? Are you in a no-win, do-or-die situation? Sometimes, God allows us to get there in order to tear us away from our comfort zone. No bird learns how to fly by staying in the warm, cozy nest constructed by its parents. At some point, we all need to learn how to take a leap of faith, kind of like Harrison Ford did in the movie Indiana Jones and the Temple of Doom.

His mission was to apprehend a particular idol from the temple of doom, and as he fled the temple with the idol in hand, Indiana found himself at the edge of a very steep cliff. He knew there were steps in front of him, but he couldn't see them, so he grabbed a hand full of dirt and threw it where he thought the steps would be, and when he did, sure enough, he could see them.

Sometimes, the path we need to take isn't clear at first. In fact, many times, it is invisible...to the natural eye, that is. God makes a way for us, but we must learn to see it by faith. The truth is that is the only way we can see God's path for our lives. The Bible makes it clear that we walk by faith and not by sight. Our carnal minds tell us that if we can see it, we can believe it. The Spirit of God within us tells us that we must first believe it before we can see it.

Hebrews chapter eleven teaches us that faith is the substance of things hoped for and the evidence of things not seen. In the natural realm, that is absurd and completely backward, but in the spiritual realm, it makes perfect sense. In the natural realm, seeing is believing, but in the spiritual realm, believing is seeing.

When Elisha, the prophet of God, was being pursued by the king of Aram, his servant Gehazi woke one morning to find the hills full of horses and chariots surrounding them. He frantically asked Elisha, "What shall we do?" "Don't be afraid," the prophet answered. "Those who are with us are more than those who are with them." And Elisha prayed, "Open his eyes, LORD, so that he may see." Then the LORD opened the servant's eyes, and he looked and saw the hills full of horses and chariots of fire all around Elisha." (2 Kings 6:16-17)

Can you see the hand of God upon your life? Do you realize that those who are with you are more than those who are with them? The enemy is always outnumbered when he stands against a child of God. Scripture teaches us that one of us can put 1,000 enemies of God to flight and two of us 10,000! Go ahead and seize the day, take a leap of faith, and believe that you and God are a majority.

Second Kings chapter seven records the story of four very desperate, leprous men who found themselves between a rock and a hard place. You see, there was a very severe famine in the land at that time, and their enemy had them surrounded on every side, just waiting for them to starve to death or surrender.

This is how they responded, "They said to each other, 'Why stay here until we die? If we say, 'We'll go into the city,'—the famine is there, and we will die. And if we stay here, we will die. So, let's go over to the camp of the Arameans and surrender. If they spare us, we live; if they kill us, then we die." It was really a no-brainer...they had nothing to lose by surrendering to their enemy because all their options would have led them to certain death.

But they chose to carpe diem. "Carpe" is a Latin word meaning "seize" in English. "Diem" means "Day". Hence, the Latin phrase "Carpe Diem" means " Seize the day " or "Seize the moment." The lepers made a unanimous decision to take a chance with their fate and go for it. It was an all-or-nothing move, and they had nothing to lose.

Chapter Ten

THE STRONGMAN

"How can anyone enter a strong man's house and carry off his possessions unless he first ties up the strong man? Then he can plunder his house."

— Matthew 12:29

O nce we understand what strongholds are and how dangerous they are, we then need to know how to break them and gain victory over them. Left unchecked, strongholds can wreak havoc in our lives and leave us defeated and destitute. So, how do we break the power of strongholds? First, we must understand that this is not a battle with flesh and blood; rather, we are fighting against the spiritual principalities and powers of the air. "For our struggle is not against flesh and blood, but against the rulers, against the authorities, against the powers of this dark world, and against the spiritual forces of evil in the heavenly realms." (Ephesians 6:12)

Since our battle is in the realm of the spirit, our weapons must be spiritual as well. "The weapons we fight with are not the weapons of the world. On the contrary, they have the divine power to demolish strongholds. We demolish arguments and every pretension that sets itself up against the knowledge of God, and we take captive every thought to make it obedient to Christ." (2 Corinthians 10:4-5). Like the expression, "fight fire with fire," believers in Jesus must learn to fight spiritual battles with spiritual weapons or face certain defeat.

No soldier goes into battle without protection of some kind. Some wear bullet-proof vests, some wear Kevlar suits, and others use camouflage to disappear from the enemy's sight. Those forms of protection are fine for physical battles and physical enemies, but soldiers enlisted in the army of the Lord are prescribed a strategic covering that is unseen by the human eye.

"Therefore, put on the full armor of God, so that when the day of evil comes, you may be able to stand your ground, and after you have done everything, to stand. Stand firm then, with the belt of truth buckled around your waist, with the breastplate of righteousness in place, and with your feet fitted with the readiness that comes from the gospel of peace. In addition to all this, take up the shield of faith, with which you can extinguish all the flaming arrows of the evil one. Take the helmet of salvation and the sword of the Spirit, which is the word of God." (Ephesians 6:13-17)

When David asked to fight with Goliath, King Saul encouraged him to wear his armor for protection. David tried to wear the armor, but it proved to be more awkward and restricting because it wasn't fit for a child. David chose not to accept the physical armor provided by Saul, but he was clothed with much more than physical armor. David was clothed in faith, and his armor was the presence of God in his life. To David, this was a spiritual battle, and he was clothed in spiritual armor. He had no doubt that he would be victorious over this giant who possessed much more experience in fighting and who towered over him at nine feet nine inches.

"He had a bronze helmet on his head and wore a coat of scale armor of bronze weighing five thousand shekels; on his legs, he wore bronze greaves, and a bronze javelin was slung on his back. His spear shaft was like a weaver's rod, and its iron point weighed six hundred shekels. His shield bearer went ahead of him." It was obvious that David was way in over his head in the natural. There was no feasible way he had any chance of victory over this mammoth man who was well-protected and had every advantage in his favor. That is, except the most important one, faith in God.

Faith proved to be exactly what David needed to gain the advantage over Goliath. David had no armor, no shield, no sword or javelin, no military experience, no armor bearer, etc. David's advantage was unseen in the natural realm but very visible in the spirit realm. David confidently ran toward Goliath, shouting, you come at me with a sword, a spear, and a javelin, but I come against you with the name of the Lord of hosts. How can a name win a battle over a sword, a spear, and a javelin? Only by faith in God who reminds us that the battle belongs to the Lord. "For everyone who has been born of God overcomes the world. And this is the victory that has overcome the world—our faith. Who is it that overcomes the world except the one who believes that Jesus is the Son of God?" (1 John 5:4–5).

Goliath represents what the bible calls the strongman. Mark 3:27 indicates, "No one can enter a strong man's house and plunder his goods unless he first binds the strong man. And then he will plunder his house." Eventually, the Israelites plundered the Philistines, but first, David dealt with the strongman, the front man whose purpose was to intimidate and discourage his enemy. Goliath was successful for forty days to intimidate the army of God and to paralyze them with fear until the man of faith came along. David's size and scope reflect that it only takes faith the size of a mustard seed to move mountains, or in this case, a giant.

I believe that God has granted us the ability to plunder our enemy, but first, we also must deal with the strongman. The devil guards our plunder with a "giant" of intimidation and fear. We can either respond like the army of Israel, with overwhelming trepidation, or we can respond in faith like David and overcome the strongman. The key is our focus; looking at Goliath results in fear and doubt, while looking at Jesus results in faith and belief. As long as Peter had his eyes on Jesus, he could walk on water, but when he shifted his focus to the wind and the waves, he began to sink.

Where is the Strongman?

Our enemy, the devil, has a place over which he seeks to control, manipulate, master, and keep hidden from the kingdom of God and God's control. There's a battle taking place against the powers of darkness and the kingdom of God. There is a spiritual realm where spiritual activity takes place, just as there is a natural realm where natural activity takes place 24/7. The difference is that you cannot see the spiritual realm, at least not with natural eyes. As we walk in the Spirit, we can discern and sense things that are taking place in the Spirit realm. The Apostle Paul reminds us in Romans 8:9, "You, however, are not in the realm of the flesh but are in the realm of the Spirit, if indeed the Spirit of God lives in you."

According to John 14:30, Satan, the strongman, is the "ruler of the world." He has also been called the god of this world, and his throne has been on planet Earth ever since the day he was kicked out of heaven. In the book of Job, we see Satan presenting himself before the Lord, indicating that he had been roaming back and forth on the earth. It doesn't take much to detect the evidence of the strongman's presence on the earth.

That is not the only place the strongman abides. He is also present among the unregenerate, "We know that we are children of God and that the whole world (unregenerate man) is under the control of the evil one" (1 John 5:19). All of humanity enters this world in original sin. Therefore, prior to our conversion to Christ, we "formerly walked according to the prince of the power of the air, of the spirit that is now working in the sons of disobedience" (Ephesians 2:2).

Bind the Strongman

To bind or defeat the Strongman, we need to understand the issue of authority. Jesus said He gave us "authority to tread on serpents and scorpions, and over all the power of the enemy" (Luke 10:19). Webster defines authority as the power to influence or command thought, opinion, or behavior. It also implies the ability to warrant or authorize an action.

This is very important information when it comes to understanding how to bind the strongman because the enemy understands authority as well. Jesus said that He would build His church, and the gates of hell would not prevail against it. Then He issued His disciples spiritual authority, "Whatever you bind on earth will be bound in heaven, and whatever you loose on earth will be loosed in heaven" (Matthew 18:18).

According to Wikipedia, binding and loosing is originally a Jewish Mishnaic phrase also mentioned in the New Testament, as well as in the Targum. In usage, to bind and to loose simply means to forbid by an indisputable authority and to permit by an indisputable authority. Originally, this custom was practiced by the Rabbis of Judaism, but Jesus has made it possible for all believers to possess this spiritual authority.

We can forbid evil spirits passage into our homes, our families, and our thoughts, and we can permit the Holy Spirit to have His way in our lives, our families, our finances, etc. We have spiritual authority over the demonic, which grants us jurisdiction over principalities and powers of darkness. When we bind the devil and lose the power of the Holy Spirit, we demonstrate the legal authority of Jesus in our lives. Jesus didn't reason with demons. He simply cast them out.

Spiritual freedom is realized by those believers who take Jesus at His Word and act in faith, exercising authority to bind and lose. Keys are a symbol of authority, but simply having a key does not demonstrate authority; rather, using the key to unlock a door is demonstrating authority. It is one thing to know that Jesus delegated spiritual authority to His flock, it's quite another thing to activate that authority and bind and loose spirits in the powerful name of Jesus. Take the advice of James, "Do not merely listen to the word, and so deceive yourselves. Do what it says." (James 1:22)

Speak the Word of God

There is power in the name of Jesus! Demons tremble at His Name. According to Scripture, there is Power in believing and in speaking the Word of God in our everyday lives, in any situation we might face. 2 Corinthians 4:13 says, "I believed and therefore I spoke," we must believe the Word of God and know that it is absolute truth, then we must speak the truth of His Word into our circumstances.

Salvation is obtained by speaking the Word. "The word is near you; it is in your mouth and in your heart," that is, the message concerning faith that we proclaim: If you declare with your mouth, "Jesus is Lord," and believe in your heart that God raised him from the dead, you will be saved. For it is with your heart that you believe and are justified, and it is with your mouth that you profess your faith and are saved." (Romans 10:8-10)

Ephesians chapter six details the armor of God that believers must wear in spiritual warfare. Part of that gear is the sword of the Spirit, which is the Word of God. The Word of God is likened to a double-edged sword, a mighty weapon that, when spoken, supernatural victory can be anticipated. When God created the earth, He spoke it into existence. When Jesus speaks, His sheep recognize Him because they know his voice.

Logos & Rhema

In Greek, which is the original language of the New Testament, two different words are used to refer to the word of God; one is *Logos*, and the other is *Rhema*. *Logos* is used to refer to the constant, written word, which is recorded in the Bible. *Rhema* is used to refer to the instant, personal speaking of God. So, the Logos is the written word, that which you read in your bible, and the Rhema is the spoken word, revealed in your spirit.

When we read the Bible, we should pray for revelation to comprehend what we read. When the Holy Spirit reveals the understanding of the scriptures, it can then manifest as a Rhema word, which is spoken to our spirit. This is why the Gospel is a stumbling block to sinners because they do not understand the meaning of scripture because it has not been revealed to them.

Guard Your House

"When a strong man, fully armed, guards his own house, his possessions are safe. But when someone stronger attacks and overpowers him, he takes away the armor in which the man trusted and divides up his plunder" (Luke 11:21–22). The devil is always looking for an open door, an opportunity to attack us. My wife is constantly asking me if I remember to lock the doors before I get into bed. As frustrating as that may be, she has a valid concern: we live in New Orleans, Louisiana, which is the murder capital of the United States!

Just as we wouldn't intentionally leave a door unlocked for a potential burglar, we shouldn't leave a door open for a spiritual attack. The devil has no authority in our lives if we guard our house, but if we knowingly or unknowingly create an invitation by leaving a window cracked open or a door unlocked, the enemy will gain access to us, and we will forfeit our spiritual authority, thus giving place to the demonic.

The Bible warns us about giving place to the enemy, "In your anger do not sin:" Do not let the sun go down while you are still angry, and do not give the devil a foothold." (Ephesians 4:26-27) The meaning of a foothold is a secure position, especially a firm basis for further progress or development. So, if we give the devil a foothold, we grant him an opportunity for future potential to wreak havoc in our lives.

You might remember the song that says, "Shut the door, keep out the devil, shut the door, keep the devil in the night, shut the door, keep out the devil, light the candle, everything's all right." It is very important for us to shut every door of opportunity that the devil may try to access. Our freedom depends on it. Satan is looking for any potential openings in our spiritual armor so that he might inflict us with his fiery darts. Therefore, take this advice from the Apostle Peter, "Be alert and of sober mind. Your enemy, the devil, prowls around like a roaring lion looking for someone to devour." (1 Peter 5:8)

Resist the Devil

Once we become aware of the schemes of the devil, we need to fight back. Peter goes on to say, "Resist him, standing firm in the faith, because you know that the family of believers throughout the world is undergoing the same kind of suffering. And the God of all grace, who called you to his eternal glory in Christ, after you have suffered a little while, will himself restore you and make you strong, firm, and steadfast." (vv. 9-10)

Webster defines resist as, to exert force in opposition, to withstand the force or effect of, and to exert oneself so as to counteract or defeat. The Bible exhorts us, "Submit yourselves, then, to God. Resist the devil, and he will flee from you." (James 4:7). The way to get the devil off your back is to resist him. Resist his temptations and do the exact opposite of what he encourages you to do. This will frustrate and thwart the plans your enemy has for you, and as he did with Jesus in the wilderness, he will leave you and come back another time.

Chapter Eleven

FAMILIAR SPIRITS

"Great spirits have always encountered violent opposition from mediocre minds."

– Albert Einstein.

T he term "As Is" implies that something is offered for sale in its present existing condition without modification. "As is" gives notice to buyers that they are taking a risk on the quality of the goods. The buyer is free to inspect the goods before purchase, but if any hidden defects are discovered after purchase, the buyer has no recourse against the seller. Any implied or express warranties that usually accompany goods for sale are excluded in an "as is" sale.

"The Spirit is God's guarantee that he will give us the inheritance he promised and that he has purchased us to be his own people. He did this so we would praise and glorify him." See 1 Corinthians 6:19-20. Before He purchased us, Jesus knew all our flaws, faults, and failures. He didn't buy us in perfect condition. He bought us AS IS! All sales final, no refunds, and no returns!

We have a Dirt-Cheap store where I live, and I sometimes like to go and hunt for bargains. They sell a wide variety of items in a wide range of conditions. Some items are missing parts, some are scratched or dented, and others are in good shape. But no matter what great bargains you may find there, nothing comes with a warranty, and nothing can be returned.

All sales are final at Dirt-Cheap, so you must really want an item, have the means to repair it, or have the missing pieces(s) at home. You can't have buyer's remorse when you shop at Dirt-Cheap because no matter how bad you feel about your purchase, you're stuck with it! You should, however, have buyer beware, which means you should take the time to examine the item before accepting it or obtain expert advice.

When Jesus purchased our salvation at Calvary, He already knew the condition we would be in, and yet He went through with the transaction, although we were broke, busted, and disgusted. 1 Corinthians 1:27-29 gives us a picture of our pre-condition; "Brothers and sisters, think of what you were when you were called. Not many of you were wise by human standards; not many were influential; not many were of noble birth. But God chose the foolish things of the world to shame the wise; God chose the weak things of the world to shame the strong. God chose the lowly things of this world and the despised things—and the things that are not—to nullify the things that are so that no one may boast before him."

That scripture tells me that God didn't call only those who had it all together, or only those who were very wealthy, or those picked most likely to succeed in High School. Rather, He chose all of us just as we are and included us regardless of our condition. God didn't choose you because you had life all figured out. He chose you in spite of it!

Why is knowing this so important? Because Satan loves to highlight our flaws and failures. His desire is to magnify our sins and to trap us in condemnation and guilt. He relentlessly slanders believers and tells them that God has buyer's remorse over them. Some believers are deceived into thinking that they are too far gone for God to "fix" them. This puts distance between believers and God and breeds discouragement and potential defeat.

What we all need to be reminded of is the fact that God is not finished with us yet. Philippians 2:13 encourages us; "For it is God who works in you to will and to act in order to fulfill his good purpose." God chooses to work in us and accomplish His perfect will, knowing that we don't have life all figured out. We must learn to shift the focus from our inadequacies and shortcomings to our all-powerful, ever-present God, who is able to do immeasurably more than we can ask or imagine.

It's not how much of Jesus you have that matters...it's how much of you that Jesus has! John the Baptist said it this way, "He must become greater; I must become less." See John 3:30. The more we decrease, the more Jesus increases in us. By surrendering more of our desires and decisions to Him, we make room for His direction and leadership to guide us.

Why is it often difficult to choose God's way? It is because of our carnal minds, the nature all humans are born with. Human nature wants its own way, not God's way. Humanly, we simply do not want to obey God. On this side of heaven, we will continually struggle to submit to God and to walk in His ways, but we must determine to surrender our will to His will if we want our lives to have meaning and impact. Paul said in the eighth chapter of Romans, "The carnal mind is enmity against God, for it is not subject to the law of God; neither indeed can it be."

One definition of the word insanity is to do the same thing over and over again and to expect a different result. If we continue doing the same things the same way, we will always get the same result. But what if we are unaware that we are stuck in that cycle? What if we don't know that we keep repeating the same actions over and over again? Perhaps we watched our parents and our grandparents do the same thing for generations and called it a tradition or a family secret.

I heard a funny story about a particular family secret that got passed down for many generations. One Christmas, a husband was watching his wife prepare a ham for their annual Christmas feast. She cut both ends off the ham and placed it into a pan, then put the pan in the oven. Her inquisitive husband asked her why she cut both ends off the ham, and the wife responded, "I don't really know. It's something my mother always did."

That answer wasn't good enough for the husband, so he asked his mother-in-law why she cut both ends of the ham before cooking it. Her answer was the same as her daughter's, "I don't really know why. It was something my mother did." So, the mother-in-law asked her mom why she cut both ends of the ham off, and she said, "Because I only had a small pan, and the whole ham wouldn't fit."

That's usually the way it goes. By the time a tradition or family secret travels a few generations, the meaning gets lost, and we don't really know why we do what we do. After generations of feuding between the Hatfields and the McCoys, no one really knows what started the feud in the first place! That is exactly how generational curses continue to exist; they stay under the radar. A great-grandparent was seduced by a particular demonic stronghold, and it remained undetected for generations, and now it is prevalent in the great-grandchildren. No one really knows how it is there, but each generation has struggled with the same demonic principality and the corresponding bondage.

Familiar Spirits

"Regard not them that have familiar spirits, neither seek after wizards, to be defiled by them: I am the Lord your God." (Leviticus 19:31). Alcohol, abuse, anger, and many other demonic strongholds are passed down from generation to generation undetected and unattested to the demise of the next generation. These wicked forces are called familiar spirits. According to Webster's dictionary, a familiar spirit is a spirit or demon that serves or prompts an individual. A demon that served a family member of a previous generation will continue to serve future generations so long as its victim is unaware.

Familiar spirits stay under the radar because their activity is confused with a bad habit or as a weakness in a person's life. The word "familiar" in the phrase familiar spirit comes from the Latin word *familiaris*. This refers to a friend or a familiar acquaintance—one who is intimately attached. This is why addictions can be so difficult to overcome, because the demonic spirit is so closely attached to the individual, like a very close friend or family member.

Pornography, gambling, overeating, etc., are spirits in disguise hoping to build a relationship with unsuspecting victims. Those things seem quite harmless at first, but after a while, they become habitual. Then, the victim becomes infatuated and, like a largemouth bass, takes the bait and swallows the hook. Now, instead of the victim calling the shots, it is the demon who calls the shots. Scripture warns us about seducing spirits, "The Spirit clearly says that in later times some will abandon the faith and follow deceiving spirits and things taught by demons." (1 Timothy 4:1)

Familiar spirits gain access to you through something or someone that you are very familiar with or comfortable with. It is usually through ignorance that we form a bond with a familiar spirit. That is why it is so important that you know who and what you are inviting into your home and your heart. The enemy comes like a Trojan horse; what seems like a gift or a blessing on the outside is loaded with the potential to completely destroy you. While God intends for us to leave an inheritance of blessings to our children, the devil insists we leave a trail of curses instead.

Breaking the Power of Familiar Spirits

The first step in breaking the power of familiar spirits is to expose them for what they are. Like the old slogan, "call a spade a spade," we must identify this as the work of the devil. We cannot call an addiction to alcohol or drugs a bad habit. On the contrary, addiction is the work of demons who wish to gain a stronghold in our lives. Rather than ignoring the root cause of generational curses, we must bring the issue to the forefront. While rehab centers are wonderful tools to help the addicted, they serve little to no purpose if they do not deal with the root cause.

I'm not certain how it happened, but I have become an avid weed-puller. My wife and I work on several flower beds, and one thing is for sure, weeds will surface sooner or later. I have found that staying on top of it makes it so much easier to weed the flower bed than letting the weeds pile up. Weeds are relentless; you have to pull up the entire root system with the weed, or it will grow back. Weeds are also very aggressive; they can take over your flower bed if allowed. Familiar spirits are like weeds. They are uninvited and unwanted and can choke the life out of that which they attach themselves to.

If we're going to break the power of familiar spirits, we must go all the way to the root of the issue. If we only address the symptoms of anger, lust, or worry, we will only see temporary results. At the root of prejudice, pride, or any other stronghold is a demon incognito, and unless that demon is exposed and addressed, its identity remains concealed. The longer the demon has been undetected, the more potential damage there is to repair. Like weeds,

119

demons like to subtly intertwine themselves around their target, with the intention of someday taking over. We will only find freedom from strongholds and addictions when we attack the root.

Jesus explains this, "The kingdom of heaven is like a man who sowed good seed in his field. But while everyone was sleeping, his enemy came and sowed weeds among the wheat and went away. When the wheat sprouted and formed heads, then the weeds also appeared." (Matthew 13:24-26) Did you notice when the enemy came and sowed weeds into the field? It was at night, in the darkness, while he was asleep. When the servants asked how the weeds got there, their master said that it was the work of an enemy. Recognizing and exposing the enemy is the first step in breaking the power of familiar spirits.

Once we recognize and expose the source, the next step is to repent and ask God to forgive us for allowing the enemy access to our hearts. "If we confess our sins, he is faithful and just and will forgive us our sins and purify us from all unrighteousness." (1 John 1:9). True repentance guarantees that we are purified and delivered from all unrighteousness, which includes demonic interference in our lives. When we repent, we do an about-face, which means we turn 180 degrees from what we have been delivered from.

Familiar spirits, like weeds, cannot thrive in an environment where they are constantly exposed and uprooted from their place. Do not give the devil a foothold in which to build a stronghold in your life, and he will back off and eventually look elsewhere to do his business.

Chapter Twelve

UNEQUALLY YOKED

"You don't have to let broken-down relationships break you down."

— Lisa Terkeurst.

Did you know that Halley's Comet is a short-period comet visible from Earth every 75–76 years? It's true; Halley is the only known short-period comet regularly visible to the naked eye from Earth and the only naked-eye comet that might appear twice in a human lifetime. You don't get many chances to see this amazing comet, so if it's on your bucket list to see, you better clear your calendar and get into the best possible position to view it because if you miss it, you may never see it again.

A wealthy businessman who was well known for being ruthless and unethical told Mark Twain that before he died, he wanted to make a pilgrimage to the Holy Land.

He said that when he got there, he wanted to climb to the top of Mount Sinai, where Moses received the Ten Commandments, and read the Ten Commandments aloud at the top. "I have a better idea," replied Twain, in his typical wit, "You could stay in Boston and keep them."

There are many things in life you don't want to miss out on, like your child's birth, graduation, and wedding day. You certainly don't want to miss your spouse's birthday or your anniversary...ask me how I know! As important as those occasions are, there is usually grace if, for some reason, you just couldn't make it. But the arrival of the Son of God foretold for centuries and clearly documented in history books...that's another thing! If you fail to recognize the arrival of the Messiah, you're in big trouble!

Jesus the Messiah arrived here in Bethlehem and lived a brief thirty-three years on the earth, and many of His own chosen people did not recognize or receive Him. John 1:11 reports the unfortunate account, "He came to that which was his own, but his own did not receive him."

Many of the Jews were looking for a heavy-handed ruler to come and overthrow the Roman empire. They had grown weary of the oppression of their overlords, and they were anticipating a savvy political type of Messiah or a rugged military type of Messiah, but certainly not a humble carpenter! Consequently, they rejected Jesus and labeled Him a common criminal. They refused to hear what He said and demanded that He be crucified. Sadly, they missed their time of visitation, and unlike Halley's Comet, Jesus would not make another visit to Earth in their lifetime.

They missed their one and only shot at recognizing the long-awaited prophetic arrival of the Messiah. This was Jesus' response to them, "For days will come upon you when your enemies will build an embankment around you, surround you and close you in on every side, and level you, and your children within you, to the ground; and they will not leave in you one stone upon another, because you did not know the time of your visitation."

Jesus exposed the lack of discernment of the Jewish people when He said, "Jerusalem, Jerusalem, you who kill the prophets and stone those sent to you, how often I have longed to gather your children together, as a hen gathers her chicks under her wings, and you were not willing. Look, your house is left to you desolate. For I tell you, you will not see me again until you say, 'Blessed is he who comes in the name of the Lord,'" (Matthew 23:37-39).

How did this happen? How could those who call themselves Shepherds of God's people not recognize Jesus as God in the flesh? Other people were able to recognize Jesus for who He was. Why couldn't these religious men and women recognize Him? I believe the reason for their spiritual blindness was the fact that they were so yoked to their religious customs and regulations that they could not embrace the radical teachings that Jesus presented. The Pharisees and the Sadducees were so engrained in their doctrine that they could not and would not embrace any teaching that differed from theirs.

A yoke is defined as a wooden bar or frame by which two draft animals (such as oxen) are joined at the heads or necks for the purpose of working together. In Hebrew culture, the word yoke was a term that was used to describe submission. So, when someone was described as yoked to someone or something, it was conveying the idea that he or she was in submission to that person or thing. In this case, the religious leaders were deeply submitted to their rituals, customs, and teachings.

People in Jesus' day readily understood analogies using a yoke. They knew what Jesus meant when He said, "Come to me, all you who are weary and burdened, and I will give you rest. Take my yoke upon you and learn from me, for I am gentle and humble in heart, and you will find rest for your souls. For my yoke is easy, and my burden is light" (Matthew 11:28–30). An "easy" yoke meant that the burden being shouldered was light and not heavy because Jesus Christ would be pulling with us. Being yoked together with Jesus means that the weight of our struggles and burdens in life is carried by Him and not us alone. Psalm 55:22 admonishes us, "Give your burdens to the Lord, and he will take care of you. He will not permit the godly to slip and fall."

To be "yoked together" is to be in a binding relationship. A marriage is a type of yoke because two individuals make a vow to become one and to forsake all others as long as they live. A contract is also similar to a yoke in that it binds two separate parties to keep the agreement specified in the contract until the contract expires. Some yokes are short-lived, while others last a lifetime. Being yoked in a marriage for a lifetime requires thorough contemplation and a strong commitment prior to making vows to do so. This is why the Apostle Paul encourages us not to become unequally yoked.

"Do not be yoked together with unbelievers. For what do righteousness and wickedness have in common? Or what fellowship can light have with darkness?" (2 Corinthians 6:14) Unfortunately, this is where so many unsuspecting believers forfeit their freedom. Forming friendships with ungodly people can seem harmless at first, but like weeds that go untouched, the enemy slowly entangles believers with corruption and darkness and chokes out their freedom. Believers become yoked to unbelievers, and soon, it becomes difficult to determine who is a believer and who is not.

The Bible further warns us, "Therefore, come out from among unbelievers, and separate yourselves from them, says the Lord. Don't touch their filthy things, and I will welcome you." (2 Corinthians 6:17). It is impossible for you to become unequally yoked with an unbeliever if you separate yourself from them. This is not to say that we utterly reject those outside of the faith. On the contrary, it is our mission to reach them, but we cannot reach them if we are yoked together with them.

Many unsuspecting Christians form unwholesome bonds with unbelievers in an attempt to win them to Christ. We are not told to completely sever ourselves from society. I believe we can even befriend the lost, but there is a fine line between maintaining a strong witness and building an unholy alliance. This is one of Satan's most successful tactics; ensnaring well-meaning Christians into a trap by fostering relationships with those he intends to yoke us with.

Years ago, my wife and I served as youth pastors, and we always had students approach us and introduce us to their boyfriend or girlfriend. Usually, the first question I asked was, "Is he or she a believer?" The typical response was "not yet," denoting the fact that they hoped to lead their new friend to Christ in the near future. Sadly, the odds were stacked against them. Rarely did we ever witness the boyfriend or girlfriend's conversion to Christianity. What usually happened was the members of our youth group slowly drifted away from the church and sometimes even drifted away from God.

Adults face the same dilemma, especially when it comes to remarriage. The odds of finding the perfect replacement for your former spouse are slim to none. The field is narrowed somewhat for divorcees, and growing older adds to the equation, so some folks become desperate to find a suitable companion. This is where compromise raises its ugly head and entices so many believers to go astray. Be careful not to justify making a poor decision especially if you have not given God an opportunity to show Himself mighty on your behalf.

There is a big difference between having an acquaintance and being yoked to someone. I believe that we should have many acquaintances with unbelievers with the idea of setting an example for them to follow. Most people come to Christ as the result of the influence of a close friend or relative. We cannot ignore the great commission which admonishes us to go into all the world and make disciples of all nations, see Matthew 28:19. We also cannot ignore James' warning, "You people are not faithful to God! You should know that loving what the world has is the same as hating God. So anyone who wants to be friends with this evil world becomes God's enemy." (James 4:4)

There is a healthy balance that we can have with unbelievers, which will encourage them to want what we have. One gauge to keep you on track is to constantly ask yourself the question, "Who's influencing who?" In addition, search the Word of God for scriptures that will keep you in check with your relationships, like this one, "Your heart should be holy and set apart for the Lord God. Always be ready to tell everyone who asks you why you believe as you do. Be gentle as you speak and show respect." (1 Peter 3:15)

One great way to prevent yourself from becoming unequally yoked with unbelievers is to surround yourself with God-fearing friends and family members who will provide an accountability partnership with you. Be yoked to God, yoke yourself to your spouse and other faith-focused believers while maintaining a healthy balance between godly relationships and those with the people of the world. Keep in mind that you are in the world, but you're not in the world.

Break the Yoke

There were many instances in the bible where the Israelites were bound by their captors, and their captors placed a heavy yoke or burden upon them, such as Pharoah did in Egypt. Pharoah burdened the Israelites with the task of building his empire, brick by brick. When he was threatened by Moses to let the Israelites go free, he made life more difficult for them and made them gather their own straw to make bricks. The Jews resented and even hated the Romans because of the Roman Empire's heavy-handed taxes and laws. One of those laws required anyone twelve years or older to carry a soldier's gear for an entire mile if they were asked. But God promises over and over in the bible to break the yoke of bondage imposed by Israel's oppressors. Like this one in Isaiah 58:6, "Is not this the fast that I choose: to loose the bonds of wickedness, to undo the straps of the yoke, to let the oppressed go free, and to break every yoke?"

Satan is a cruel taskmaster. He loves to yoke unsuspecting victims to accomplish his wicked schemes. His aim is much like that of Pharoah's and Caesar's, to burden others with the task of building his kingdom. These captives lack discernment and have no earthly idea that they are contributing to Satan's schemes because his cunning and deceptive ways easily hide the truth from them. For example, the same devil who entices and tempts someone to sin then yokes that person with a burden of guilt or shame when they succumb to the temptation. He catches them with the hook of temptation, but he binds them on the yoke of guilt and shame. However, God has promised to break this type of yoke as well. "Come to me, all who labor and are heavily laden, and I will give you rest. Take my yoke upon you, and learn from me, for I am gentle and lowly in heart, and you will find rest for your souls. For my yoke is easy, and my burden is light." (Matthew 11:28-30)

Galatians 5:1 summarizes the theme of this book, "It is for freedom that Christ has set us free. Stand firm, then, and do not let yourselves be burdened again by a yoke of slavery." Once we have tasted freedom from the bondage of the enemy, we never want to fall into that trap again, but the devil, on the other hand, believes that if he fooled you once, he could fool you again. It's best to avoid areas of vulnerability and weakness. If you have been delivered from alcoholism, don't let anyone talk you into operating a nightclub ministry. No doubt you have a burden for those folks because you can identify with their struggle. However, unless you hear an audible voice from heaven or see writing on the wall, it is wise to steer clear of bars, taverns, and nightclubs.

Once we are delivered from the clutches of the enemy's grip, God intends for us to stay free. One of the realities of life that truly puzzles me is the issue of repeat offenders. There are those who knowingly break the law, believing that it is a necessary evil. They get caught and thrown in prison for a particular sentence. The purpose of the sentence in prison is to encourage the prisoner to make better choices in the future. However, statistics show that recidivism rates in the U.S. are some of the highest in the world, with almost 44% of criminals released returning to prison within their first year out. You would think that one visit to prison would be sufficient to bring any criminal into reform, but apparently, it is not.

Unfortunately, recidivism rates for spiritual offenders are just as high, if not higher, than for criminal offenders. Once the newness of our deliverance from bondage to a particular area of sin wears off, demons try to entice us to become repeat offenders. Jesus taught that once a demon has been cast out of an individual, it comes back later to check on the situation. If the delivered person has not filled the vacancy in their heart with the presence of Jesus, that demon goes and finds other demons who come and torment the person, and now his situation is far worse than it was before Jesus set him free!

The way to stay free is to break the yoke, and in case you are wondering, you can't do it by yourself. So many have tried in vain to break the yoke and to lift their heavy burdens. Only God can break the yoke of bondage as He did many times for Israel. "Now I will break the yoke of bondage from your neck and tear off the chains of Assyrian oppression." (Nahum 1:13). Notice who is breaking the yoke in this passage; it is God who breaks our yoke and sets us free from the chains of our oppressor.

Here are some ways that you can stay free from the yoke of bondage:

Live a life of repentance – asking God to forgive your sin is one thing, but repenting of your sin is totally different. Many repeat offenders never get past asking God to forgive them. God will forgive you every time you ask Him to, but why do you keep falling into the same trap? Because you have not repented. Repentance means to choose to turn in the opposite direction. Forgiveness is what God does. Repentance is what you and I do. "Repent, then, and turn to God, so that your sins may be wiped out, that times of refreshing may come from the Lord." (Acts 3:19)

Live with accountability – If you are struggling with an area of sin and you want to be free, invite a godly person who demonstrates a life of faith and has proven to be trustworthy to keep you accountable for your actions. "My brothers, if anyone among you wanders from the truth and someone brings him back, let him know that whoever brings back a sinner from his wandering will save his soul from death and will cover a multitude of sins." (James 5:19-20)

Pray without ceasing – If you want to live free from bondage, prayer must become a lifestyle for you. God promised those who seek Him with all their hearts will find Him. "Rejoice always, pray continually, give thanks in all circumstances; for this is God's will for you in Christ Jesus." (1 Thessalonians 5:16-18)

Stay in the Word – God's Word is a lamp for our feet and a light for our path. It is our spiritual road map; we cannot navigate our journey of faith without it. It takes the power of the Word to break demonic strongholds and the bondage of yokes. "All Scripture is God-breathed and is useful for teaching, rebuking, correcting, and training in righteousness, so that the servant of God may be thoroughly equipped for every good work." (2 Timothy 3:16-17)

Out Of Egypt

PART FOUR

FREE AT LAST!

Chapter Thirteen

IDENTITY IN CHRIST

"The more you reaffirm who you are in Christ, the more your behavior will begin to reflect your true identity."

— Neil T. Anderson

A s a pastor, I have the responsibility to oversee the way the sheep of my flock walk out of their faith. For the most part, I am very encouraged by their behavior. They possess a genuine hunger for God's presence and seek Him wholeheartedly in worship. Many of them utilize their gifts and talents to serve in ministry. They shockingly stay focused while I preach, lol, and even "amen" me frequently. They show great kindness and appreciation by showering us with compliments and gifts on many occasions. My wife and I are truly blessed to shepherd the flock that God has assigned to us.

However, as is typical with so many Christians, I often compare them to a drunk driver weaving all over the road; they start in the Spirit one moment, and the next moment, they swerve off the road into the flesh. The longer they stay in the flesh, the more prone they are to end up in a ditch. If they can manage to stay in the Spirit, they can safely navigate the will of God for their lives.

The key to staying in the Spirit involves the same principle we must apply in order to drive safely: keep your eyes on the road! Since the invention of cell phones, we have more distracted drivers than ever before. Rather than watching the road, many drivers watch their phones and become reckless and dangerous to those around them. To stay in the Spirit, we must keep our eyes on Jesus and refuse the many distractions around us that cause so many to veer off the road of faith.

"Very truly, I tell you, Pharisees, anyone who does not enter the sheep pen by the gate but climbs in by some other way is a thief and a robber. The one who enters by the gate is the shepherd of the sheep. The gatekeeper opens the gate for him, and the sheep listen to his voice. He calls his own sheep by name and leads them out. When he has brought out all his own, he goes on ahead of them, and his sheep follow him because they know his voice. But they will never follow a stranger; in fact, they will run away from him because they do not recognize a stranger's voice." Jesus used this figure of speech, but the Pharisees did not understand what he was telling them." (John 15:1-6)

Why do you suppose the Pharisees did not understand Jesus' parables? Were they deaf or hard of hearing?

Perhaps, but I think it was more likely that they were not on the same page with Jesus. They couldn't identify with Him. They had selective hearing, and consequently, what Jesus said made little sense to them. They only heard what they wanted to hear, and they discarded the things that revealed their spiritual pride and unmasked their obvious arrogance.

The question is not whether God speaks to us on a regular basis. On the contrary, I believe that He constantly speaks to us. The real issue is whether or not we listen to Him. If you're struggling to hear His voice, remember that God speaks in a still, small voice. You must be quiet and attentive if you want to hear Him speak. He created us with two ears and only one mouth for a reason.

Tune out all distractions and make time to learn to listen to His still, small voice. As you learn to identify His voice, you will also learn to discern the enemy's voice. Unfortunately, the devil likes to communicate with us, too. Satan likes to disguise his voice to confuse and confound us with his deception and lies. In time, we will recognize that he is the voice of a stranger, and we will learn to avoid it entirely.

Knowing Who & Whose We Are

"For if we live, we live to the Lord, and if we die, we die to the Lord. So then, whether we live or whether we die, we are the Lord's." (Romans 14:8). Scripture talks about God the Holy Spirit in many ways in the Old and New Testaments, but simply put, we experience the Holy Spirit as the presence and power of God. After Jesus' resurrection, the Holy Spirit has become "God with us." Everything we experience in our faith comes through the work of the Holy Spirit—in us, through us, and with us!

So, how would you live differently if, at any moment, you knew you could access the power of God's Spirit? Our hope is that together, we discover that not only is this possible, but it is what God created us for! The Holy Spirit is sometimes called the "master sculptor," chiseling and chipping away everything in our lives that doesn't look like Jesus. The Spirit is the engine of power and transformation for the Christian life.

The Apostle Paul had this in mind when he said in Galatians 5:25, "If we live by the Spirit, let us also keep in step with the Spirit." Are you regularly in touch with the Holy Spirit? On this side of heaven, it is imperative that we walk in unity with and stay in touch with the Holy Spirit as He guides us. Since we walk by faith and not by sight, it is through revelation from the Spirit of God that we understand our course.

It stands to reason that if we choose to ignore the leading of the Spirit of God, we will find ourselves distant from His presence and out of touch with His will for our lives. Paul continues," Those who live according to the flesh have their minds set on what the flesh desires; but those who live in accordance with the Spirit have their minds set on what the Spirit desires. The mind governed by the flesh is death, but the mind governed by the Spirit is life and peace. The mind governed by the flesh is hostile to God; it does not submit to God's law, nor can it do so. Those who are in the realm of the flesh cannot please God." (Romans 8:5-8)

Our Identity in Christ

Webster defines identity as the distinguishing character or personality of an individual. At the core of what it means to be a Christ follower is to receive a new identity. In Jesus, we do not lose our true selves, but eventually, we become our true selves only in Him. In Christ, we are fundamentally new and belong to the family of God. The language values, customs, and expectations of this world have increasingly become foreign to us. We have been born again for another world, to a greater kind of existence.

"For all of you who were baptized into Christ have clothed yourselves with Christ. There is neither Jew nor Gentile, neither slave nor free, nor is there male and female, for you are all one in Christ Jesus." (Galatians 3:27-28) Kenneth Hagin said, "One of the most important revelations we can get from the Word of God is to understand who we are in Christ. Identifying with Christ will change the way we live and cause us to rise above adversity. Not understanding our identity in Him will keep us living far below our rights and privileges in Christ." Life is not about finding yourself; it is about discovering who God created you to be.

The new identity that you receive in Christ far outweighs any other identity you may have. Your gender, your family, your age, race, culture, school, career, accomplishments, and titles may all reveal very important aspects about you, but they are not who you are. Your identity is rooted in Christ and in His character. You are a new creation that never existed before; the old you is dead, and the new you is alive forevermore. Allow me to point out several truths about your new identity in Christ:

You are loved. (1 Thessalonians 1:4)

You are chosen. (Ephesians 1:4)

You are accepted (Romans 15:7)

You are redeemed (Isaiah 43:1)

You are forgiven (Colossians 2:13)

You are empowered (2 Peter 1:3)

You are gifted (Romans 12:6)

You are blessed (Ephesians 1:3)

Identity Theft

Identity theft is a crime in which a hacker uses fraud or deception to obtain personal or sensitive information from a victim and misuses it to act in the victim's name, which is why we have so many usernames and passwords to keep up with! Sounds like the work of the devil to me! Satan uses fraud and/or deception to obtain personal or sensitive information from believers so he can use it against them. The devil knows the power we wield as believers when we recognize who we are in Christ. The Bible declares that we are overcomers in Christ, more than conquerors, heirs of God, and joint heirs with Jesus Christ!

Our personal relationship with God is the foundation for everything in life. God is the source of our identity and our purpose in Christ. It is He who created us, and it is He who defines us. We are not defined by our circumstances, our jobs, our possessions, or our position in society. We are solely defined by who we are in Christ. Knowing our identity in Christ is of utmost importance because the enemy seeks to steal our identity. There is a constant war taking place for the identity of believers. "Be sober-minded; be watchful. Your adversary, the devil, prowls around like a roaring lion, seeking someone to devour. Resist him, standing firm in your faith." (1 Peter 5:8-9)

Knowing our identity in Christ is critical because we behave based on what we believe about who we are. Distorted belief about our identity leads us to distorted behavior. If we have a false belief about who we are, we behave and act in ways contrary to God's perfect will for our lives. Far too many believers suffer from poor self-esteem because they believe the lies of the enemy about who they are. Their identity has been stolen, and they are clueless about it. We are who God says we are, not who the world says we are, not who our co-workers say we are, and certainly not who the devil says we are.

All human beings – male and female – are created in the image of God. Each is a beloved son or daughter of heaven, and as such, each has a divine destiny. Most believers don't see that there is a direct connection between the way they view themselves and their God-given destiny. You are not how you feel. You are not your past mistakes; you are not your current frustrations. You are unique and one of a kind. You are the handiwork of God, His creation, and the result of His love for you. Sadly, we are too quick to forget these truths and too eager to believe the lies that Satan pours out day after day. This is our battle for spiritual freedom. We cannot afford to back down. God is for us, so who can be against us?

Some of the most profound examples of our identity in God come from the book of Exodus. When God summoned Moses to deliver the Israelites from the bondage of Egypt, He demonstrated many supernatural signs and wonders. The plagues that God sent to the Egyptians were a stark reminder of the distinction between the two nations. For example, Moses told Pharaoh that "if you refuse to let them go and continue to hold them back, the hand of the Lord will bring a terrible plague on your livestock in the field – on your horses, donkeys, and camels and on your cattle, sheep, and goats. But the Lord will make a distinction between the livestock of Israel and that of Egypt so that no animal belonging to the Israelites will die." (Exodus 9:2-4)

All the horrible plagues that came upon Egypt had zero impact on the people of Israel. God shielded His children and protected them because He identified with them as their God. Perhaps the most profound example was the last plague in which God destroyed all the firstborn males in Egypt, both men and animals. God told Moses that there would be loud wailing throughout Egypt – worse than there ever was or ever will be again, but among the Israelites not a dog will bark at any person or animal. Then you will know that the Lord makes a distinction between Egypt and Israel." (Exodus 11:6-7) We will focus in more detail on distinction in the next chapter, but it is imperative that we, as believers, understand that we are set apart and highly favored as the children of God.

"For you are a people holy to the Lord your God. Out of all the peoples on the face of the earth, the Lord has chosen you to be his treasured possession." (Deuteronomy 14:2) The Lord chose Israel, and the Lord has chosen you. The Bible is chocked full of admonitions of love from the Lord to His people. "But you are a chosen people, a royal priesthood, a holy nation, God's special possession, that you may declare the praises of him who called you out of darkness into his wonderful light." (1 Peter 2:9)

Mistaken Identity

According to Webster's Dictionary, mistaken identity is defined as a situation in which someone or something is mistakenly thought to be someone or something else. Christians often use the term "hypocrite" when referring to someone who calls themself a Christian, but their actions say otherwise. Webster defines a hypocrite as a person who acts in contradiction to his or her stated beliefs or feelings. Let's take a look at some possible causes for mistaken identity, according to the International Christian Coaching Institute.

Abuse – verbal, emotional, physical, and sexual abuse in the home, at school (bullying), or in the workplace.

Comparison/favoritism – growing up with parents who played favorites or constantly compared you to your siblings.

Cultural Influences – social media, movies, celebrities, fashion trends, and cultural narratives (i.e., the "self-made" man, the American Dream, etc.); how well you perceive yourself to fit in with your culture.

Faulty thinking – having a negative or pessimistic view of yourself, believing lies about yourself, negative or self-condemning self-talk, having a victim mentality, regularly attributing bad circumstances to personal factors rather than those outside your control.

Major life transitions – entering a new school, graduating, career changes, getting married, becoming a parent, children moving out, getting a divorce, losing a loved one, retirement, and many other major life transitions.

Physical issues – deformities, handicaps, illnesses, injuries, size or weight issues.

Success or failure – how you respond to success and failure; how much you attribute success and failure to yourself or to circumstances.

Unrealistic expectations – the standards you set for yourself or the expectations others have of you – and how well you meet or fail to meet those standards.

With so many factors to influence your identity and self-image, it's easy to drift off course and not understand who God created you to be. That's why it's crucial to know God's Word and follow His leading so your life and identity have a firm foundation. Ephesians 4:22-24 is a guide for understanding our new identity, "You were taught, with regard to your former way of life, to put off your old self, which is being corrupted by its deceitful desires; to be made new in the attitude of your minds; and to put on the new self, created to be like God in true righteousness and holiness." You are not defined by your past or by what other people say about you, and certainly not what the devil says about you.

For the sake of our witness as a child of God, we cannot allow anything to hinder our focus. The list of causes above represents many but not all of life's "curve balls" that can be thrown at us. Rather than striking out all the time, I say we learn how to hit a curve ball. Stay in the batter's box, don't count yourself out, and take a swing at whatever it is that is challenging your faith. Who knows, you might just hit it out of the park for a home run! And when life stacks up the "bases," it just turns a homerun into a grand slam!

Counterfeit Christians do harm to the body of Christ. The Bible says, "All you need to say is simply 'Yes' or 'No'; anything beyond this comes from the evil one." (Matthew 5:37). Let your words be few but honest. Our walk must line up with our talk if we are going to let our light shine before others. Our example is all that some people will ever have to decide whether or not to follow Jesus, and quite possibly all they will ever need. It is vital that we pay close attention to our attitude and our actions as it pertains to those watching us on a regular basis. 1 John 2:6 warns us, "Whoever claims to live in him must live as Jesus did."

What are some things we can do to safeguard our witness? How can we ensure that we are making a positive impact on those around us? Here are a few ways we can let our light shine:

Seek God's will, not your own. (John 6:38)

Pray with compassion. (Hebrews 5:7)

Work for God, not men. (Colossians 3:23)

Daily deny yourself. (Luke 9:23)

Be willing to suffer. (1 Peter 2:21)

Give thanks continually. (1 Thessalonians 5:18)

There is no greater guarantee that we will be a positive witness than when we mimic the original source Himself. "Imitate God, therefore, in everything you do because you are his dear children. Live a life filled with love, following the example of Christ. He loved us and offered himself as a sacrifice for us, a pleasing aroma to God." (Ephesians 5:1-2) The Apostle Paul challenged the church at Corinth to follow his example as he followed the example of Christ. That was a bold challenge given the fact that Paul was once an enemy of the gospel and heavily persecuted the church. But it also signifies that once you receive a new identity in Christ, you too can encourage others to follow you as you follow Christ.

Chapter Fourteen

OUR DISTINCTION

"Why work so hard to fit in when you were called to be set apart?"

- Unknown

Out of all the nations on the planet, God chose the tiny nation of Israel to be His chosen people. The scriptures tell us, "You are a people holy to the Lord your God. The Lord has chosen you out of all the peoples on the face of the earth to be his people, his treasured possession." (Deuteronomy 7:8). The fact that the Jews are God's chosen people means that they have been held to a high standard, for those who are given much, much is required. The Israelites were set apart, which means to separate something and keep it for a special purpose or to make someone seem distinctive or superior.

The word distinction implies being separate, different, or distinguished from others. So, it could be said that God's people are set apart from the world for His special purposes. While that distinction is a great honor, the Israelites didn't always see it that way. At times, they chose rather to become like the people around them, to live like they lived, to worship like they worshipped, and to think like they thought. God gave Moses the Law, which was a strict code of conduct to which the Israelites were to adhere, without compromise. As long as they were obedient and lived according to the Law, the Israelites were blessed, and they prospered in every way. However, when they turned their backs on God and disobeyed the Law, they faced hardship and great difficulty.

In extreme cases, God would allow Israel's enemies to prevail upon them, plunder them, and, in some cases, carry them into exile. Five times in their history, the Israelites were exiled to a foreign country for their sin and rebellion against God.

- The first exile took place in Egypt from 1523 BC – 1313 BC.
- The second exile took place in Babylon from 423 BC – 372 BC.
- The third exile took place in Persia/Media from 372 BC – 348 BC.
- The fourth exile took place in Greece from 371 BC – 140 BC.
- The fifth exile took place in Rome, which began in 69 BC and has not yet ended.

Despite all of the persecution and discipline that they endured on behalf of their sinful behavior, the Israelites found a way each time to return to their God in repentance and humility. In the words of Mark Twain: "The Egyptians, the Babylonians, and the Persians rose, filled the planet with sound and splendor, then faded to dreamstuff and passed away; the Greeks and Romans followed and made a vast noise, and they were gone…The Jew saw them all, survived them all."

A Remnant Remains

God had and will always have a remnant. A remnant is defined as that which is left of a community after it undergoes a catastrophe. God's remnant is the people who acknowledge Him in all their ways, even when their ways sometimes do not please Him. God's remnant has a distinct confidence in His presence, a strong hope in hardships, they trust in His authority, and they possess an optimism concerning the future. Ezra 9:8 reveals this, "But now, for a brief moment, the Lord our God has been gracious in leaving us a remnant and giving us a firm place in his sanctuary, and so our God gives light to our eyes and a little relief in our bondage."

Noah and his family were a very small remnant saved out of the millions of people who inhabited the earth prior to the flood. From the tiny seed of eight people, God replenished and repopulated the earth. Following the showdown on Mount Carmel with Jezebel and the prophets of Baal, the prophet Elijah despaired that he was the only one left in Israel who had not bowed down to worship idols, but God assured him that He had reserved a remnant of 7,000 who had not bowed to worship Baal. In the sermon on the mount, Jesus spoke of the remnant that would inherit eternal life, "Enter through the narrow gate. For wide is the gate, and broad is the road that leads to destruction, and many enter through it. But small is the gate and narrow the road that leads to life, and only a few find it." (Matthew 7:13-14)

The prophet Isaiah spoke of this centuries ago when he said, "Though the number of the Israelites be like the sand by the sea, only the remnant will be saved. For the Lord will carry out his sentence on earth with speed and finality." The final remnant will be comprised of any and all who call upon the name of the Lord. Father God desires that none would perish and that all would come to repentance. The fact that a remnant has always remained of the people of God is attributed to the grace and mercy of Jehovah, who is described as long-suffering, slow to anger, and abounding in love.

Psalm 30:1-5 provides us with a picture of God's unrelenting love for His people; "I will exalt you, Lord, for you rescued me. You refused to let my enemies triumph over me. O Lord my God, I cried to you for help, and you restored my health. You brought me up from the grave, O Lord. You kept me from falling into the pit of death. Sing to the Lord, all you godly ones! Praise his holy name. For his anger lasts only a moment, but his favor lasts a lifetime! Weeping may endure through the night, but joy comes with the morning."

If you are a born-again Christian, you are a very distinct individual. You have become an heir of God and a joint heir with His Son, Jesus Christ. Your name has been recorded in the Lamb's Book of Life, and you have the promise that Jesus is preparing a place for you in heaven. And if that is not enough, at the end of your life in this world, you will hear your heavenly Father say to you, "Well done, good and faithful servant! You have been faithful with a few things; I will put you in charge of many things. Come and share your master's happiness!" (Matthew 25:23) For all eternity you will walk on streets of gold, never shed a tear, never be afraid, and never see darkness.

Set Apart

From his birth, Moses was set apart unto God for a special purpose. When Pharoah issued an edict in Egypt to destroy all the Hebrew male children, Moses' parents hid him in a reed basket and placed it in the Nile River because they took notice that he was a special baby and wanted to keep him from being destroyed by the enemy. Moses was found by Pharoah's daughter, who also recognized something special about him. She decided to keep him, and she unknowingly placed the child in the care of his mother, Jochebed, who was also paid to take care of the child. What an awesome God we serve! He not only provided protection from Pharoah, but He also gave him right back to his mother, who was paid to raise him with privilege. You can't make that stuff up!

God knew that approximately eighty years later, Moses would be set apart even further to become the deliverer for His chosen people from their bondage in Egypt. From day one, God had His hand upon Moses. Somehow, the devil knows when a deliverer is about to emerge. He attempted to kill Jesus in His infancy as well. "Then Herod, when he saw that he had been tricked by the wise men, became furious, and he sent and killed all the male children in Bethlehem and in all that region who were two years old or under, according to the time that he had ascertained from the wise men." (Matthew 2:16-18)

Your new nature in Christ has set you apart from the world. This is the way the Bible describes you, "But ye are a chosen generation, a royal priesthood, a holy nation, a peculiar people; that ye should show forth the praises of him who has called you out of darkness into his marvelous light." (1 Peter 2:9) Like Moses and Jesus, we have been set apart as deliverers to rescue those who are held in bondage to the enemy. That is why it is so important for us to walk in obedience to the Word of God and to keep ourselves from becoming tainted by the ways of the world because we cannot allow ourselves to be caught in a snare of bondage. The devil targets deliverers, and captives are counting on us to stand in the gap and intervene as a deliverer.

God's presence in our lives distinguishes us from unbelievers. There have been many times when, simply by looking into someone's eyes, I could tell that they were a child of God. Something about their countenance made it evident that they belonged to the Lord. A strong sense of peace and contentment was obvious on their face, as well as confidence in God's presence. Moses also knew how important the presence of God was in relationship to the task God had given him. "Then Moses said to him, "If your presence does not go with us, do not send us up from here. How will anyone know that you are pleased with me and with your people unless you go with us? What else will distinguish me and your people from all the other people on the face of the earth?" (Exodus 33:15)

A Different Spirit

What does it mean to have a different spirit? It means that you possess the resilience to go against the flow if necessary. It means that you are not concerned about popular opinion as it relates to your character and integrity. It also means that you are a God-pleaser and not a man-pleaser when it comes down to it. Caleb was a man who possessed a different spirit. God told Moses that not one of the Israelites who saw His glory and the signs He performed in Egypt and in the desert would ever see the land He promised them except for Caleb. "But because my servant Caleb has a different spirit and follows me whole-heartedly, I will bring him into the land he went to, and his descendants will inherit it." (Numbers 14:24)

A different spirit is an opposite spirit. It works much like the kingdom of God…in reverse! The Bible instructs us that to become great, we must become a servant; to be first, we must be last, and to gain our lives in the kingdom, we must be willing to lose them. Furthermore, in times of famine, we share. In times of danger, we serve. In times of poverty, we give. In times of adversity, we sing. In times of stress, we rest. In pools of hatred, we love. In times of persecution, we bless. When assaulted, we forgive.

Qualities of a Different Spirit

- **Aggressive attitude** – "Then Caleb quieted the people before Moses and said, "Let us go up at once and take possession, for we are well able to overcome it." (Numbers 13:30)

- **Optimistic outlook** – "But Joshua the son of Nun and Caleb the son of Jephunneh, who were among those who had spied out the land, tore their clothes; and they spoke to all the congregation of the children of Israel, saying; "The land we passed through to spy out is an exceedingly good land." (Numbers 14:6-7)

- **Tenacious trust** – "If the Lord delights in us, then He will bring us into this land and give it to us, a land which flows with milk and honey." (Numbers 14:8)

- **Fearless faith** – "Only do not rebel against the Lord, nor fear the people of the land, for they are our bread; their protection has departed from them, and the Lord is with us. Do not fear them." (Numbers 14:9)

Do you recall any of these names from the Bible...Shammua, Shaphat, Igal, Palti, Gaddiel, Gaddi, Ammiel, Sethur, Nahvi, and Geuel? I didn't think so. That's because they failed in their mission to bring back a positive report from spying out the land of Canaan. These men were chosen leaders among the tribes of Israel, but they were common, ordinary people. They did not possess a different spirit like Joshua and Caleb, who brought back a glowing report from Canaan and confirmed what God told Moses, that it was a fruitful and fertile land of great promise. Those ten men were denied access, along with an entire generation of Israelites, from entering Canaan because of their lack of faith and confidence in God. In fact, they were all struck down and died of a plague.

Caleb's life and circumstances prove that there can be no compromise or retirement when it comes to doing the work of the Lord. At the tender age of eighty-five, Caleb returned to the border of what would soon be the Land of Israel. When the Lord's challenge was issued to him in his old age, he was still eager to do His will, "Now then, just as the Lord promised, he has kept me alive for forty-five years since the time he said this to Moses, while Israel moved about in the wilderness. So here I am today, eighty-five years old! I am still as strong today as the day Moses sent me out; I'm just as vigorous to go out to battle now as I was then. Now, give me this hill country that the Lord promised me that day. You yourself heard then that the Anakites were there, and their cities were large and fortified, but the Lord helping me, I will drive them out just as he said." Then Joshua blessed Caleb, son of Jephunneh, and gave him Hebron as his inheritance." (Joshua 14:10-13).

The promises of God are yes and amen, but it requires a different spirit to recognize and inherit those promises. Many times, those wonderful promises are met with staunch resistance from both our peers as well as our adversaries. Satan's goal is to drive fear into our hearts and tempt us to question God's faithfulness and ability. He also desires to bring a gulf of division between the people of God. But, when God finds a person with a different spirit, He will fulfill His promise to him or her no

matter how long it takes. So, at eighty-five, God gave
Caleb the strength to overcome the same giants he had
seen forty-five years earlier and inherit their land.

Light vs. Darkness

Born-again believers are a distinct breed of people.
They are not of this world; their residence is within the
kingdom of God on earth. Their marching orders come
from 2 Corinthians 6:17, "Come out from them and be
separate, says the Lord. Touch no unclean thing, and I will
receive you." Those who have wholeheartedly committed
their lives to serving the Lord Jesus are like a fish out of
water as it relates to their lives in this world.

"For now, we see through a glass, darkly; but then
face to face: now I know in part; but then shall I know
even as also I am known." (1 Corinthians 13:12). The Bible
likens our lives as a mist or a vapor that is here today and
gone tomorrow. Compared to eternity, which is never-
ending, the span of our lifetime pales in comparison to
that of eternal life. No matter how hard a believer tries to
fit in with the world, there is a marked difference between the saved and the unsaved. Paul told the Corinthian church that when someone is born again, their old life
is gone, and a brand-new life has begun. That person will
never be the same again; he or she is a new creation with
a unique identity in Christ.

Matthew 5:16 encourages us to let our light shine
brightly in this dark world, "Let your light so shine before
men, that they may see your good works, and glorify your
Father which is in heaven." Christians have a different
spirit than those of the world because the Spirit of God

has come to live in their hearts. Believers are the temple of the Holy Spirit, the carriage that carries King Jesus!

Chapter Fifteen

SPIRITUAL ORDER

"A good man's parking space is ordered by the Lord."

— Paul H. Palser.

Our Heavenly Father is a God of order. He created everything in an orderly sequence in a six-day span that set the world as we know it into motion. He created the sun, moon, and stars to regulate time and seasons, and the heavenly bodies operate with predictable precision. Then He created Adam in His own image, and from Adam, He created Eve, who was to be a suitable companion for Adam. He placed them in the Garden of Eden and gave them the responsibility to work it and to take care of it. God made all kinds of trees to grow in the garden that were both pleasing to the eye and good for food. God made a river to water the garden that flowed from Eden and separated it into four tributaries.

161

Everything God created had structure, form, and order. There was no chaos in the garden, and there was absolute harmony and beauty all around Adam and Eve. Adam and his wife were both naked, and they felt no shame. Before He created Eve, God told Adam, "You are free to eat from any tree in the garden, but you must not eat from the tree of the knowledge of good and evil, for when you eat from it, you will certainly die." (Genesis 2:16-17)

Demonic Disorder

With all of the splendor and beauty of Eden and the very presence of God to accompany them each day, how in the world was it possible for Adam and Eve to be deceived and to rebel against God? Two reasons: first, God gave them free will to choose, and second, the devil was present in the garden. Satan cannot stand spiritual order because it honors God and creates an atmosphere for the blessing and favor of the Lord toward those who walk in obedience to His commands. Satan saw all that God had done for Adam and Eve and how they were blessed in every way, and he couldn't stand it.

"Now the serpent was craftier than any of the wild animals the Lord God had made. He said to the woman, "Did God really say, 'You must not eat from any tree in the garden?" This is how all spiritual disorder is initiated; Satan twists the words of God in an effort to confuse and deceive his victims. If he is successful, the devil can thwart the plans of God for our lives and cause great pain and suffering. God's Word is the foundation for spiritual order, and it is the cornerstone on which our faith stands. "For the word of God is alive and active. Sharper than any double-edged sword, it penetrates even to dividing soul

162

and spirit, joints and marrow; it judges the thoughts and attitudes of the heart." (Hebrews 4:12)

Unfortunately, many Christians do not fully understand spiritual order, and because of their lack of knowledge in this area, they needlessly suffer lack. The kingdom of God is a spiritual kingdom, and those of us who are born of the Spirit of God have been called to navigate within the spiritual order that governs His spiritual kingdom. My family and I like to travel, and as anyone who has traveled to an unknown destination knows, if you do not have a good set of directions, a reliable map, and a common understanding of how to read and follow a set of directions, chances are that you will get lost and suffer the loss of time and resources.

I begrudgingly admit that I am directionally challenged, so it is a good thing that my wife and daughter are skilled at finding the best routes and giving directions on our trips. It is the same principle as we travel on our spiritual journey; we are traveling to an unknown destination, unknown to us, that is because God knows exactly where He is taking us. So, if we do not understand the spiritual order that governs our journey, we will struggle to navigate the will of God for our lives. That is why we are commanded that in all things, we should strive to follow the path of spiritual order that God has foreordained. "But everything should be done in a fitting and orderly way." (1 Corinthians 14:40)

Divine Alignment

If, while driving your vehicle, you notice that your wheels gradually drift from the center of the road or pull in one direction or the other, it might be time for a front-end alignment. A trained mechanic can assess the problem and properly align your vehicle so that you can once again drive your vehicle safely. People of faith experience a similar experience; over time, we are pulled in one direction or another, and sometimes, we drift from the center and weave all over the road of life. Fortunately, we have a "spiritual mechanic" whose name is Jesus, who can assess our problems and re-align our hearts and minds so that we can remain on the narrow path that leads to life.

"The steps of a good man are ordered by the Lord, and He delights in his way." (Psalm 37:23). My father-in-law loved this verse of scripture. Whenever I would ride with him to the store, and he found a parking spot close to the entrance, he would say, "A good man's parking spots are ordered by the Lord." Now, I find myself saying the same thing whenever I get a good parking spot. When we are in divine alignment, God is able to lead us where He wants us to be. Some people call being in the right place at the right time luck, and I call it divine alignment.

When our lives are out of divine alignment, we may experience a range of emotions such as negativity, fatigue, loss of motivation, and spiritual, mental, and physical burnout. There may even be times when we feel like God has abandoned us because we are not in divine alignment anymore. Life's circumstances seem too difficult to handle, the pressure of life becomes too great to endure, and we fail to seek comfort from the One who has overcome the world, conquered sin and death, and has all power in His hands. We begin to look for answers in other things only to realize they are leading us down the wrong path, and we are heading for potential destruction.

God's Positioning System

According to Wikipedia, "The Global Positioning System (GPS), originally Navstar GPS, is a satellite-based radio navigation system owned by the United States government and operated by the United States Space Force. It is one of the global navigation satellite systems (GNSS) that provides geolocation and time information to a GPS receiver anywhere on or near the earth where there is an unobstructed line of sight to four or more GPS satellites." This technology provides us with precise directions in which to navigate anywhere in the world.

As wonderful as The Global Positioning System is, there is another system that is even greater than that. I call it God's Positioning System. God's Positioning System provides Christians with precise directions in which to navigate our spiritual journey. It keeps us on the straight and narrow path, and it leads us in the way that we should go. So, how does God's Positioning System work? Jesus explains it like this, "...my sheep hear my voice, and I call my own sheep by name, and I lead them out. And when I bring out my own sheep, I go before them; and the sheep follow me, for they know my voice." (John 10:3-4)

You might say that God's Positioning System is voice-activated. We know which way to go because, in our spirit, we hear the voice of our Shepherd leading us. Isaiah describes it this way, "Whether you turn to the right or to the left, your ears will hear a voice behind you, saying, "This is the way; walk in it." (Isaiah 30:21) Some would say, "God doesn't speak to me, I don't hear His voice." In my personal experience, I believe God speaks to all of His children, but unfortunately, not all of His children are listening. When God revealed Himself to Elijah, He told him to stand on the mountain, for He was about to pass by. First, a strong wind came and shattered the rocks, but the Lord was not in the wind. Next, there was an earthquake, but the Lord was not in the earthquake. Then a fire, but the Lord was not in the fire. Finally, a still, small voice revealed the presence of God to Elijah.

God works in mysterious ways, and He speaks in various ways through various circumstances. Many people are disappointed when God doesn't show up when and how they expected Him to show up. Perhaps too many of us are expecting God to show up like a mighty rushing wind, an earthquake, or a raging fire, and we miss our opportunity because God chooses to speak in a still, small voice. God is not limited, nor will He be confined to our carnal ideals when He chooses to reveal Himself to us. We must constantly be reminded that our ways are not God's ways, and our thoughts are not God's thoughts. His ways and His thoughts are much higher than ours.

Your Will Your Way

Naaman was the commander of the army of the king of Aram. He was a great man in the sight of his master and highly regarded, and he was a valiant soldier, but he had leprosy. He was urged to go to Elisha, the prophet,

166

and ask him to heal his leprosy. With anticipation, Naaman approached Elisha and presented his need, to which Elisha told him that if he dipped seven times in the Jordan River, he would be healed. Upon hearing this, Naaman became angry and said to Elisha, "I thought that he would surely come out to me and stand and call on the name of the Lord his God, wave his hand over the spot, and cure me of my leprosy. Are not Abana and Pharpar, the rivers of Damascus, better than all the waters of Israel? Couldn't I wash in them and be cleansed?" So he turned and went off in a rage." (2 Kings 5:11-12)

Fortunately for Naaman's sake, his servant talked some sense into his head, and he reconsidered Elisha's orders. He dipped seven times in the Jordan, and his skin became clean like that of a young boy. I wonder how many other people could've been healed of a sickness or delivered from a stronghold had they, like Naaman, humbled themselves and responded according to what God told them to do. It's not enough to agree to do God's will. We must also agree to do it His way. I pray this phrase often, "Help me to do your will your way, Yahweh."

King Saul learned this lesson the hard way. This is the word God sent to Saul, "I will punish the Amalekites for what they did to Israel when they waylaid them as they came up from Egypt. Now go, attack the Amalekites and totally destroy all that belongs to them. Do not spare them; put to death men and women, children and infants, cattle and sheep, camels and donkeys." (1 Samuel 15:2-3) It was God's will to avenge His children for what the enemy had done to them by totally annihilating them, but Saul did God's will his way, and he paid dearly for it. Because Saul kept King Agag as a prisoner and the best of the livestock for sacrifices, God sent Samuel to Saul and told him that He regretted making him king and that the kingdom would be taken from him for his disobedience and given to David.

Kingdom Order

The kingdom of God is governed by the Word of God and functions with biblical order. God established the five-fold ministry of Apostles, Prophets, Evangelists, Pastors, and Teachers to equip His people to do the work of the ministry. Healthy, productive believers know their place in the kingdom and operate under the divine order of God and His Word. This obedience provides a covering of God's grace and blessing in which to navigate the will of God and to faithfully accomplish His purposes. Jesus told His disciples, "The student is not above the teacher, nor a servant above his master. It is enough for students to be like their teachers, and servants like their masters." (Matthew 10:24-25) The key to living in kingdom order is to know your place and to stay in your lane.

Prophet, Priest, and King

Disorder breeds chaos and calamity. We cannot allow the pressures of life and dire circumstances to break kingdom order. When the Philistines gathered to make war with the Israelites, King Saul became terrified when he saw the vast army assembling before him. Saul's fear spread to his troops, who were also terrified. The size of the Philistine army was not the only concern Saul had; the Israelites did not have proper weapons to fight with because no blacksmith could be found in all the land of Israel. The only weapons Saul's soldiers had were plowshares, mattocks, axes, and sickles. Many of Saul's army had already hidden themselves in caves, thickets, holes, and pits, and to make matters worse, the prophet Samuel was not present to authorize the sacrifice and to secure the blessing of the Lord for battle. So, Saul took matters into his own hands and offered the sacrifice himself, which turned out to be a huge mistake.

The proper kingdom order would've been for the prophet to inquire of the Lord on behalf of the king, and if there was to be a sacrifice, the priest would do the honors; then, and only then, would the king be authorized to go into battle. Obviously, Saul allowed his immediate circumstances and his dire straits to override his convictions, and he willfully broke kingdom order, which invited chaos into the camp of the Lord. We must guard against making the same mistake ourselves; no matter how difficult or urgent our situation may be, we must strive to keep kingdom order our highest priority.

Family Matters

If our homes are to be places of peace and protection, kingdom order must be a priority. The Bible gives us clear instructions of God's intended order within our families. Ephesians 5:21-23 explains, "Submit to one another out of reverence for Christ. Wives, submit yourselves to your own husbands as you do to the Lord. For the husband is the head of the wife as Christ is the head of the church, his body, of which he is the Savior." Paul urges husbands to love their wives as Christ loves the church, and in chapter six, he continues by admonishing children to obey their parents in the Lord, for this is right. He further encourages them to honor their father and mother, which is the first commandment with a promise, so that it may go well with them, and the reward is that they may enjoy long life on the earth. That, in a nutshell, is how kingdom order works within the family, and when it is honored, it produces blessing and favor. For many years, families in this country have been under siege by the constant attacks of Satan and his demonic hoards.

Far too many households are fatherless, and many homes are riddled with physical, sexual, and verbal abuse. What results is confused and troubled children, where total disorder and chaos fill the house. Family after family continues to be plundered by an unseen enemy who has unholy intentions to kill, steal, and destroy any and all potential for a healthy, happy home. Nearly one-half of all marriages in this country end in divorce, causing unprecedented division and disorder within our society. Substance abuse, violence, confusion, anger, and frustration fill our streets largely due to the breach of kingdom order within the home. When the home is out of order, the community will be out of order. When the community is out of order, society will be out of order, and eventually, an entire nation will be void of biblical order, and anarchy and chaos take over.

The Freedom Cycle

According to Alexis de Tocqueville, a French aristocrat and historian, the average age of the world's greatest civilizations from the beginning of history has been about 200 years. During those 200 years, these nations always progressed through the following sequence: From bondage to spiritual faith; From spiritual faith to great courage;

From courage to liberty; From liberty to abundance; From abundance to selfishness; From selfishness to complacency; From complacency to apathy; From apathy to dependence; From dependence back into bondage.

Our great nation began with a strong religious impetus when the Pilgrims landed in 1620. Over the next century and a half, more and more people flocked to our shores seeking religious freedom. That religious quest generated tremendous courage within the British colonies. General Washington rallied a group of farmers, and we defeated the mighty British army. With that victory, we gained our independence. That being said, unfortunately, we are now at the other spectrum of the freedom cycle. The freedom cycle, in many ways, mimics the sin cycle in the Bible.

The Sin Cycle

This is how the cycle of sin unfolded in the Old Testament, "Whenever the Lord raised up a judge for them, he was with the judge and saved them out of the hands of their enemies as long as the judge lived; for the Lord had compassion on them as they groaned under those who oppressed and afflicted them. But when the judge died, the people returned to ways even more corrupt than those of their fathers, following other gods and serving and worshiping them. They refused to give up their evil practices and stubborn ways." (Judges 2:18-19)

This is how the sin cycle operates in the New Testament, "When tempted, no one should say, "God is tempting me." For God cannot be tempted by evil, nor does he tempt anyone, but each person is tempted when they are dragged away by their own evil desire and enticed. Then, after desire has conceived, it gives birth to sin; and sin, when it is full-grown, gives birth to death." (James 1:13-15) There are four stages of the sin cycle that keep its victims coming back for more. Let's examine each stage and give a brief description of it:

- **Stage One:** *DESIRE* (v. 14)

We are tempted by what we desire. Hunters trap animals by baiting the trap with what an animal desires. Satan baits traps for us based on our sinful desires, and we are lured and enticed by those very desires.

- **Stage Two:** *DECEPTION* (v. 15)

Desire is the bait on the hook, and if we take the bait, the devil sets the hook, and then we are deceived into believing that what we're about to do is harmless.

- **Stage Three:** *DISOBEDIENCE* (v. 15)

Once the hook is in our mouth, we are led away into disobedience. At this stage, we are held captive by the enemy because our sinful behavior has led us to rebel against God's commands.

- **Stage Four:** *DEATH* (v. 15)

Unrepentant sin results in spiritual death. Sin kills our joy, our peace, and our victory in Christ. The wages of sin is death, according to Romans 6:23, but the gift of God is eternal life.

It all starts with desire. That's how it started in the Garden of Eden, and that's how it starts with us. We must learn to discern what our flesh craves and then realize that is exactly where the enemy will attack. Many great men and women of the Bible were seduced and then succumbed to their own desires:

- Eve desired the forbidden fruit.
- Samson desired Delilah.
- David desired Bathsheba.
- Jezebel desired authority.

There is good news: we can overcome our desires and resist temptation. Where sin abounds, grace abounds much more. In other words, God provides ample reinforcements when we are enticed by the enemy. He always makes a way where there seems to be no way. Take refuge in the words of the Apostle Paul, "No temptation has overtaken you but such as is common to man; and God is faithful, who will not allow you to be tempted beyond what you are able, but with the temptation will provide

Out Of Egypt

Chapter Sixteen

BRACE FOR IMPACT!

"The greatest setback can lead to your greatest breakthrough. Trust and believe that all things are working together for your good."

– Germany Kent.

Spiritually speaking, the world is a very dark place, and the darkness is increasing at an alarming rate. The Bible warns us, **"But mark this: There will be terrible times in the last days. People will be lovers of themselves, lovers of money, boastful, proud, abusive, disobedient to their parents, ungrateful, unholy, without love, unforgiving, slanderous, without self-control, brutal, not lovers of the good, treacherous, rash, conceited, lovers of pleasure rather than lovers of God—having a form of godliness but denying its power. Have nothing to do with such people."** (2 Timothy 3:1-5) That pretty much describes the world we live in today. We are certainly living in the last days, and the signs of the times are all around us.

Prophetic warnings are in abundance throughout the Bible. The book of Matthew records this warning; "Nation will rise against nation, and kingdom against kingdom. There will be famines and earthquakes in various places. All these are the beginning of birth pains. Then you will be handed over to be persecuted and put to death, and you will be hated by all nations because of me. At that time many will turn away from the faith and will betray and hate each other, and many false prophets will appear and deceive many people. Because of the increase of wickedness, the love of most will grow cold." (Matthew 24:7-12)

If there was ever a time the church needed a breakthrough, it is now. "When you are tempted to give up, your breakthrough is probably just around the corner." (Joyce Meyer) As a pastor, I can testify that many churches are going through it! So many congregations suffered setbacks from the recent pandemic and have never recovered. Some churches have merged with another congregation, while others were forced to shut their doors. God's plan for His church is not for it to break up. Rather, it is for the church to break through the many barriers standing in the way. And His plan is the same for you and me. The devil is planning a breakup, but God is determined that we will have a breakthrough!

Lord of the Breakthrough

King David experienced great breakthroughs in his military campaigns for Israel. When he was newly anointed King of Israel, the Philistines came out to attack him in the Valley of Rephaim. David inquired of the Lord, "Shall I attack the Philistines? Will you deliver them into my hands?" (2 Samuel 5:19) The Lord assured David that he would succeed in the battle, so he went to Baal Perazim and there he defeated the Philistines. What does the name *Baal-perazim* mean? Well, the word *Baal* just means "Lord," and the word *perazim* means "breach" or "breakthrough." So, Baal-perazim means "Lord of the Breakthrough." That is exactly why David named the place where God broke through his enemies "Baal-perazim." David said, "As waters break out, the Lord has broken out against my enemies before me." (v. 20)

Are you facing a major battle in your life right now? Has your enemy come to attack you and plunder you? If so, you can use the same strategy that David used to ensure your victory and your personal breakthrough. Let's break down the strategy David used to ensure his victory:

Prayer – "David inquired of the Lord, Shall I go and attack the Philistines?" (2 Samuel 5:19) When the enemy threatens us, it is crucial that we seek the counsel of God. David wanted the blessing of the Lord before he went into battle because victory depended upon it, and the same is true for you and me.

Pursuit – "So David went to Baal Perazim, and there he defeated them." (v. 20) David faced his enemy with courage because he had the backing of his God. Once we have prayed and gained the favor of the Lord, action is required to ensure victory.

Praise – Once the breakthrough was won, David gave credit where credit was due. He took no credit and gave all the glory to God when he said, "As waters break out, the Lord has broken out against my enemies before me." (v. 20)

The Lord of the Breakthrough is no respecter of persons, which means what He did for David, He will do for you. Don't panic when your enemy attacks; pray instead and ask God how to proceed. If He assures you of victory, don't freeze up. Pursue the enemy head-on with confidence because if God is for you, and He is, who can be against you? And don't forget to praise and magnify God for your victory. Remember, the battle belongs to the Lord. He will fight your battles for you if you let Him. Deflecting the praise back to God indicates that you understand who is in charge and who is watching over you 24/7.

Let My People Go

After Moses faced a mammoth struggle with Pharoah, which lasted several months in Egypt, a breakthrough finally came! Following the last plague, which claimed the lives of all the firstborn males in Egypt, Pharaoh reluctantly released Moses and the Israelites from their bondage. Shouts of jubilation rang throughout the land of Goshen as the Israelites plundered their enemy and packed their bags for departure. Their suffering had ended, and their chains were broken: no more slavery, no more Pharoah, no more pyramids. The Israelites only had to leave on their minds.

What a breakthrough moment! Over four hundred years of bondage led to a monumental, supernatural breakthrough for the people of God. Deliverance from all that held them captive was now a reality, and they were completely free...at least from Pharaoh and Egypt. As wonderful as breakthroughs are, they have a way of revealing hidden truths that we bury deep down inside. During the time of the plagues, God was doing as much in the hearts of the Israelites as He was in Pharaoh's heart. This was the response of the people of Israel when they found out that Pharaoh had changed his mind and came after them, "Is it because there were no graves in Egypt that you have taken us away to die in the desert? What have you done to us in bringing us out of Egypt? Did we not tell you in Egypt, 'Leave us alone and let us serve the Egyptians'? It would have been better for us to serve the Egyptians than to die here." (Exodus 14:11-12)

That doesn't sound like a grateful response to God's miraculous deliverance of His chosen people from a harsh taskmaster. But unfortunately, it doesn't take long for the devil to steal the praise from our lips. The spirit of gratitude was dismissed by the spirit of fear, and Moses went from a hero to a zero in a very short time. What the children of Israel didn't know was that getting free is only the beginning of the process; learning how to stay free is the key to a breakthrough.

Walk in the Spirit

So, how do we always keep the proper perspective regardless of our circumstances? How can we avoid the traps of the enemy intended to ensnare us in rebellion and ingratitude? Fortunately, the Bible has an answer, and it is found in Galatians 5:16-18, "So I say, walk by the Spirit, and you will not gratify the desires of the flesh. For the flesh desires what is contrary to the Spirit, and the Spirit what is contrary to the flesh. They are in conflict with each other, so you are not to do whatever you want. But if you are led by the Spirit, you are not under the law." Walking in the Spirit is critical to maintaining our breakthrough. As you know, the enemy doesn't stop attacking when a breakthrough comes. He just looks for another opportunity to infiltrate.

"Flesh" is not our old sin nature because that died with Christ on the cross. Our "flesh" is anything one has— body, mind, emotions, patterns, beliefs, influences, etc.— that is outside of the divine resources of the Holy Spirit. When we depend on these "fleshly" resources, we are living independently from God. Walking in the Spirit, on the other hand, is choosing to live dependently upon God moment by moment and to obey His Word to the best of our ability. I want to share with you a simple way to stay in the Spirit that derives from a very important strategy that firefighters teach us if we catch on fire: stop, drop, and roll.

Stop – The most important step in walking in the Spirit is to stop walking in sin. Our old ways need to stop, our bad habits need to stop, our old thoughts need to stop, and it might be necessary for some old relationships to stop. Before we can begin with a clean slate, we need to wipe the old slate clean. "No one who abides in him keeps on sinning; no one who keeps on sinning has either seen him or known him." (1 John 3:6)

Drop – Letting go can be very difficult, but it is absolutely necessary if we are going to walk successfully in the Spirit. So, drop the attitude, drop the need to be right, drop the association with carnal living, and drop the need to be needed. Hebrews 12:1 encourages us, "Therefore, since we are surrounded by such a great cloud of witnesses, let us throw off everything that hinders and the sin that so easily entangles."

Roll – Learn how to roll with the Spirit. "The wind blows wherever it pleases. You hear its sound, but you cannot tell where it comes from or where it is going. So it is with everyone born of the Spirit." (John 3:8). In order to stay in the Spirit, we need to learn how to flow with the Spirit.

Be Prepared to Make Adjustments

I remember a time when I was in Bible college, and we were at a leadership retreat. As the student body president, I was asked to give a short talk to the leaders, and as hard as I tried, I couldn't come up with anything that I felt would be helpful. So, I asked a friend to give me something to share, and he gave me the following nugget. The title was Be Flexible. Then, two important points followed. The first point was don't be a stick in the mud, which means don't be so rigid that you won't budge for anyone or anything. When pressure comes, be like a palm tree and learn to bend and not break. And the second point was don't be like a chameleon that changes with the scenery. That means don't bend so far for others that you compromise your own convictions.

One of the best lessons we can learn in life is that when things go wrong, we can make adjustments. On a regular basis, we make many adjustments in life; we adjust our clocks for daylight saving, we adjust our schedules for interruptions, we adjust the mirrors on our vehicles, we adjust the volume on the radio, and we adjust the position of our car seats. Those things come relatively easily, but why is it so difficult to adjust our attitude? According to Merriam-Webster, the word "adjust" means to bring to a more satisfactory state, to rectify, or to adapt. My pastor has a saying when something doesn't go the way he planned, he says, "I just make a new decision."

How do we know when adjustments are necessary in life? Try this simple test: get in your car and find a road that is straight for a long way. Be sure you're driving the speed limit, then take your hands off the steering wheel and see how long it takes for your vehicle to veer one way or the other. If your vehicle is correctly aligned, it will take longer to veer than it will if it is out of alignment. If it veers to one side quickly, you need an alignment. A mechanic will hook your vehicle up to a computer which will inform him which adjustments need to happen to give your vehicle the proper alignment.

The same principle applies to believers. As we travel the straight and narrow path, how quickly we veer to the left or the right is an indicator of the condition of our alignment. I make it a regular prayer request, asking God to keep me in divine alignment with His will. I know how easy it is for me to get bent out of shape. The enemy has a good idea of what it takes to provoke us simply by observing our everyday actions. If we are going to navigate from bondage to breakthrough, we need to be willing to make adjustments along the way. Those Israelites from the Exodus who were inflexible and refused to make adjustments died in the desert. So many of them became a "stick in the mud" and refused to bend at God's command, so, eventually, they broke!

Has anyone ever told you, "You need an attitude adjustment?" Some things are easy to adjust, like adjusting the temperature on the thermostat or adjusting a ball cap to fit more comfortably, but other things are not so easy. It proves to be harder to adjust an antenna for better TV reception or to adjust the settings on your latest electronic device (I hand them over to my children to program). I will go even further and say that adjusting people can be way more difficult than adjusting things. Most of us resist change however and whenever possible, but if we are going to embrace our breakthrough and continue to live in freedom, change is inevitable.

Flexibility is a requirement for continued breakthroughs. The word "flexible" means yielding to influence. It is characterized by a ready capability to adapt to new, different, or changing requirements. (Merriam-Webster) Palm trees are amazingly flexible. Although palms may look delicate, they are actually much stronger than your average tree. They have been known to withstand winds up to 145 miles per hour, and while hurricanes uproot other trees and destroy buildings, palm trees are usually unscathed. They have the potential to bend 40-50 degrees without snapping. Palms also have lots of small roots, which spread out into the top layers of the ground, that hold on to more soil and give it a much heavier base that helps the trees stand upright.

Sustain Your Breakthrough

Because of the tremendous effort that is required to get and maintain a breakthrough, it is of great importance that we understand how to sustain our breakthrough. Prior to promotion, we must be willing to face many struggles that are specifically designed to provide us with the necessary means to champion our breakthrough. "Consider it pure joy, my brothers and sisters, whenever you face trials of many kinds because you know that the testing of your faith produces perseverance. Let perseverance finish its work so that you may be mature and complete, not lacking anything. (James 1:2-4) Sustaining a breakthrough is not possible without testing and trials. God is not interested in providing temporary victories and momentary breakthroughs. He wants us to enjoy lasting blessings with peace and joy.

God blesses what has been tested and proven in our lives. Promotion and breakthroughs come out of our times of testing and the battles we face and win. Our greatest battles prove to become our greatest breakthroughs, and the very assignments that the enemy has had against us will become our most joyful victories. "No weapon formed against you shall prosper, and every tongue which rises against you in judgment you shall condemn. This is the heritage of the servants of the Lord, and their righteousness is from me, "says the Lord." (Isaiah 54:17) Are you feeling pressure and resistance? Is the enemy attacking on all sides with relentless bombardments? If so, brace yourself for impact and get ready to witness the power of God. He has promised to fight our battles and to bring us victory if we endure the battle and trust Him to overcome the enemy.

The Power of Brokenness

For us to maintain and sustain our breakthrough, we must learn to embrace brokenness, which is godly sorrow. It is vital to understand the value of humility and the fear of the Lord if we want to continue celebrating our victories. Much like the olive is crushed to procure its precious oil, we, too must be broken so that God can get out of us the valuable gems He puts into us. A good example of this is found in Genesis 32:24-28, "So Jacob was left alone, and a man wrestled with him till daybreak. When the man saw that he could not overpower him, he touched the socket of Jacob's hip so that his hip was wrenched as he wrestled with the man. Then the man said, "Let me go, for it is daybreak." But Jacob replied, "I will not let you go unless you bless me." The man asked him, "What is your name?" "Jacob," he answered. Then the man said, "Your name will no longer be Jacob, but Israel because you have struggled with God and with humans and have overcome." I

want you to focus on some very important words from these short verses of scripture:

Alone – Jacob was singled out; he was set apart because God had an agenda for him. When you feel singled out from the crowd and all alone, it could be an indicator that God has some important business with you to accomplish.

Wrestled – Jacob wrestled all night with an angel of the Lord. He struggled with God's agenda for his life. Fortunately, God will not allow us to keep our distorted views of Him or His ways, so He allows us to wrestle it out with Him.

Wrenched – Jacob's hip was wrenched, which means it was twisted in the struggle and possibly pulled out of the socket. Jacob walked with a limp as a reminder of the incident so he wouldn't forget the significance of that encounter with God.

Daybreak – The night represents the season of struggle prior to breakthrough, sometimes referred to as the dark night of the soul. That season is different for all of us; for some, it is brief, but for others, it can seem like a lifetime. For Noah, it was around 100 years. For Abraham, it was 25 years, and for Joseph, it was 13 years. Daybreak represents a breakthrough. That was when Jacob's struggle with the angel ended. "For his anger lasts only a moment, but his favor lasts a lifetime; weeping may stay for the night, but rejoicing comes in the morning." (Psalm 30:5)

Israel – God not only changed Jacob's name, but He also changed Jacob's life that night. When God is about to do extraordinary things in a person's life, He often changes their name. He changed Abram's name to Abraham, who became the father of many nations, and He changed Simon's name to Peter, who became the pillar of the church.

One of the best examples of brokenness in the Bible is David's prayer of repentance in Psalm 51. David was grieved by his sinful behavior in seducing Bathsheba into fornication and for having her husband killed in battle. This psalm is his plea for God's mercy and grace to forgive and restore the joy of salvation. Verse 17 indicates that David understood brokenness. "My sacrifice, O God, is a broken spirit; a broken and contrite heart you, God, will not despise."

We know that God both desires brokenness and rewards broken people with the privilege of His presence and the assurance of His deliverance. We know this because Psalm 34:18 reveals, "The Lord is close to the brokenhearted and saves those who are crushed in spirit." My pastor tells me all the time not to trust a man of God who doesn't walk with a "limp." Of course, the limp he is referring to is the one Jacob acquired while wrestling with the angel, and the implication is that brokenness leaves us with physical, emotional, or spiritual scars or wounds to remind us of our desperate need for God.

Final Thoughts

FINAL THOUGHTS

Living in bondage is a wretched existence where one feels trapped and hopeless. Getting into it is much easier than getting out, and unfortunately, some never get out. According to dictionary.com, bondage is defined as "the state of being bound by or subjected to some external power or control." Much like the children of Israel, life for us is a cycle of getting caught in a trap of the devil, crying out to God for deliverance, walking in freedom, and then getting ensnared all over again. It's easier to see the cycle in the life of a person we call an "addict," which refers to someone addicted to a life-controlling substance such as alcohol or drugs. But all of us enter the cycle at one time or another. Our addiction may not be life-controlling or life-threatening, but it can still be considered bondage.

For example, you might be addicted to sweets, gossip, scrolling on your phone, shopping, or binge-watching your favorite shows. Those seem like harmless addictions, but they have a way of demanding something from us. They demand our time, our energy, our affection, and our resources, to name a few. Texting and driving is a terrible addiction, one of which I have personally become a victim of. Not long ago, a young lady sped through a red light and slammed into my car and totaled it. When she came to check on me, she asked if I was okay and then proceeded to explain that when she looked up, she realized that the light was red, but it was too late to stop before slamming into me.

I knew right away why she was looking down; it was because she was texting while driving. There are all types of bondage we get ourselves mixed up in dead-end relationships, pornography, racism, jealousy, bitterness, anger, over-eating, and the list goes on. Human nature is plagued with the desire to rebel against God and His Word. From the beginning of time, Satan has made it his ambition to keep humans in one form of bondage or another. Living in bondage is a choice we personally make. There may be lots of coercion, but in most cases, no one forces us into bondage. We choose it on our own. I realize there are cases where the victim has no choice and is overpowered by their captor. Those who are sexually, physically, or verbally abused are often victims of those who have some type of dominion or authority over them.

John 10:10 says, "The thief comes only to steal and kill and destroy; I have come that they may have life and have it to the full." Bondage is a thief that steals our joy, kills our potential, and destroys our hope. But Jesus comes to give us life more abundantly, a full and joyful life free from the entanglements of sin. Jesus came to set the captives free, and He has, and He does, but getting free and staying free are two different things. My sincere intention in writing this book is to offer biblical advice and experiential knowledge for readers to maintain their breakthroughs and stay free from the entanglements of bondage from the past. As you know, the cycle of bondage to breakthrough seemed never-ending for the Israelites. Repeatedly, they found themselves entangled in idol worship and indulging in pagan religion. Then, either God would punish them and send an army to attack them and carry them away into exile, or they would cry out to God for deliverance, and He would forgive them and restore them, but it wouldn't be long for the cycle to repeat itself.

Get to the Root

In an article by James Owen from National Geographic, studies show that most large, captive-bred carnivores die if returned to their natural habitat. The odds of animals such as tigers and wolves surviving freedom are only 33 percent, according to a team of researchers from the University of Exeter in the United Kingdom. Many animals and humans struggle to embrace freedom if they have been held captive for a long period of time. For example, if you trap a raccoon in a wire cage, he will walk back and forth in the cage and eventually realize that he is trapped. After a few days of this, even when you open the door of the trap, the raccoon will not escape because his brain has been conditioned to believe there is no way out.

Sadly, many humans find themselves in the same pitiful state, feeling trapped with no way of escape. Some feel caught in a dying marriage. Others are incarcerated by a gripping addiction, while many others see no way out of their current debt. The reason why those situations seem so hopeless is because the answers are not on the surface, you have to get to the root of the problem in order to overcome it. Just like pulling weeds in your garden, if you just pluck what you can see, you will be free of the weeds temporarily. When it comes back, and it will, if you just pluck the surface again, you will do this repeatedly. That is why many victims of abuse keep returning to their abuser because they are led to believe that they are the ones to blame and that they deserve what is happening to them.

However, if you dig down a little bit and get hold of the roots of that "weed" (the lie, deception, persuasion, etc.) and pull the roots out of the ground, you will kill the potential of its return forever. That is the message of Out of Egypt, to navigate from bondage to breakthrough by getting to the root of the issue that got you into bondage in the first place. If you do not address the root issues, you can never be totally free. Just like a dog chasing its tail, you will spin around and around chasing freedom but never achieve it if you don't get to the root cause of your bondage. It takes a little extra time and a little more effort, but the results will make or break you. The truth is, you never know how close you are to breakthrough if you give up too soon.

Just Keep Swimming

Do you remember the little fish named Dory from the movie Finding Nemo? There is a scene in the movie when Nemo gets lost, and his dad, Marlin, and his friend Dory set out to find him. At a rather discouraging point in the search, Dory attempts to encourage Marlin by saying, "Hey, Mr. Grumpy Gills, when life gets you down, do you want to know what you gotta do? Just keep swimming, just keep swimming, swimming, swimming." I would say just keep digging, digging, digging until you get to that nasty root that holds you captive. Your spouse is not the problem, your job is not the problem, and your church and/or your pastor is not your problem. The devil is your problem, and he has planted a seed deep down inside of you, and it has manifested among the very people and things that are dear to you.

Seek God to reveal what the enemy has planted in you, and ask Jesus to yank that weed from the roots and pluck it out for good! Then you will be able to quote this scripture from the heart, "So if the Son makes you free, then you are unquestionably free." (John 8:36). Being un-questionably free means that you are no longer attached to your bondage. You no longer return to your cage or to your captor again. Rather, you celebrate your freedom every day because you have been set free once and for all. That is what real breakthrough is all about, and that is God's ultimate plan for you and me: to live in freedom every day and to reflect His perfect love to those around us.

The prophet Jeremiah penned these now famous words hundreds of years ago, "For I know the plans I have for you," declares the Lord, "plans to prosper you and not to harm you, plans to give you hope and a future." (Jeremiah 29:11) The context of this declaration from God was a promise of restoration to come in spite of the sinful be-havior of Israel. God had Nebuchadnezzar carry nearly all of the Israelites into exile from Jerusalem to Babylon. He did this because He loved them, and He wanted them to have a change of heart while in exile. God spoke of the breakthrough before their bondage ever began. He already has your breakthrough planned out, too. In the midst of your captivity, God is planning your breakthrough.

There was a man whose life was characterized by failure. At the age of 22, his business failed. At 23, he ran for the Legislature but was defeated. So, he turned to another business, which failed when he was 24. At 25, things started to turn around when he was elected to the Legislature, but at 26, his love died. He had a nervous breakdown at 27 and was defeated in his run for Speaker at 29. He was defeated at 31 when he ran for Elector. He lost in his run for Congress at 34 but then managed to win at 37. He was defeated in his re-election for Congress in 39. He was defeated in his run for Senate at 46. He was defeated in his run for Vice President at 47. He was defeated in his run for Senate at 49. This man was stubborn as a mule but not a winning mule.

He was defeated over and over and over again until, at the age of 51, his breakthrough came, and he was elected President of the United States of America. In fact, he would become one of the greatest presidents in our country's history. He is Abraham Lincoln. He endured the hardships, the trials, the defeats, the rejections, and the failures until one day he became our commander-in-chief. That is the power of perseverance, and it is vital for achieving a breakthrough. Stay the course, keep on believing, and don't look back...your breakthrough is coming!

www.ingramcontent.com/pod-product-compliance
Lightning Source LLC
Chambersburg PA
CBHW040852210326
41597CB00029B/4808